The Toolbox Dialogue
Initiative

The Toolbox Dialogue Initiative

Initiative

The Power of Cross-Disciplinary Practice

Edited by
Graham Hubbs
Michael O'Rourke
Steven Hecht Orzack

CRC Press
Taylor & Francis Group
Boca Raton London New York

CRC Press is an imprint of the
Taylor & Francis Group, an **informa** business

First edition published 2021
by CRC Press
6000 Broken Sound Parkway NW, Suite 300, Boca Raton, FL 33487-2742
and by CRC Press
2 Park Square, Milton Park, Abingdon, Oxon, OX14 4RN

Library of Congress Cataloging-in-Publication Data
A catalog record has been requested for this book

ISBN: 978-1-138-34173-9 (hbk)
ISBN: 978-1-138-34168-5 (pbk)
ISBN: 978-0-429-44001-4 (ebk)

Typeset in Times
by Newgen Publishing UK

Contents

Chapter 9   Enhancing Cross-Disciplinary Science through Philosophical Dialogue: Evidence of Improved Group Metacognition for Effective Collaboration

Brian Robinson and Chad Gonnerman

Chapter 10   Qualitative Analyses of the Effectiveness of Toolbox Dialogues

Marisa A. Rinkus and Michael O'Rourke

*Sanford D. Eigenbrode, Stephanie E. Vasko, Anna Malavisi,
Bethany K. Laursen, and Michael O'Rourke*

Preface

The Toolbox Dialogue Initiative or TDI (http://tdi.msu.edu/) runs philosophically based, capacity-building workshops for cross-disciplinary research and practice teams. Over the course of a workshop—typically two to three hours—members of such teams explore the implicit beliefs and values that influence their project contributions, with the goal of improving the performance of the team. This book presents the history of TDI, its theoretical background, challenges, achievements, and impact.

As of this writing, TDI has run more than 300 Toolbox workshops around the world. The first workshops were held in order to help graduate students working in research teams at a small land grant university. In recent years, the Toolbox approach has become an internationally recognized way of training cross-disciplinary research and practice teams. TDI is now a consortium of investigators who are dedicated to using research to build the capacity and productivity of many different kinds of teams.

The idea of a book chronicling the life and times of TDI was proposed by one of the co-editors, Steven Orzack, who is a TDI advisory board member. At first, the book was to be a short, single-authored monograph written to provide the reader with all the information they would need to run a Toolbox workshop. While the contents of this book provide the reader with that information, it is now a much more collective and expansive endeavor that has contributions from many members of the TDI community. This reflects the core TDI value of collaboration. We practice what we preach in these pages.

The editors and authors of this book believe that it accomplishes a number of objectives. We describe the theoretical background of our facilitative work and provide evidence for its effectiveness. We also describe our history and offer a glimpse into our future. We present a detailed description of a dialogue method that can be used to facilitate communication and collaboration among members of teams engaged in almost any kind of project. This work is grounded in concepts and results from philosophy, communication theory, interdisciplinary theory, and psychology. Most importantly, the book provides a model for those who wish to enhance collaborative, cross-disciplinary research and practice. We know of no other accounts of such efforts that provide the conceptual and practical details that we provide here. These details are invaluable for those who wish to engage in this type of work.

This book is best read from the beginning. After describing what TDI is and what it has been, we discuss its theoretical background, best practices for running Toolbox workshops, evidence for its effectiveness, and future directions. We close with case studies of several workshops. Although it is an edited book, its chapters build on each other. That said, in Chapter 1 we describe various ways in which the book can be sampled by those who only need part of the story, for example the theoretical background, or the logistical details of running a Toolbox workshop.

The structure of this book was determined collectively, with the authors deciding how to divide TDI's story into chapters. The resulting book is somewhere between an edited volume and a collectively authored monograph. In editing it, we have made decisions that ensure consistency across the chapters. For example, "TDI" denotes the name of the research community, whereas "Toolbox" indicates elements of the workshop intervention. In addition, as we develop in more detail in Chapter 1, "cross-disciplinary" is used as a cover term for various, more specific forms of practice that combine disciplines, such as "interdisciplinarity" and "transdisciplinarity." This standardization makes for a more coherent reading experience.

We hope that this volume will help you collaborate better!

Graham Hubbs
Michael O'Rourke
Steven Hecht Orzack

Acknowledgments

We would like to thank the many people who have made TDI possible.

First, we thank two members of our advisory board—Kara Hall and Julie Thompson Klein—and current members of TDI who are not co-authors—Edgar Cardenas, Ike Iyioke, and Chet McLeskey. We have been assisted by a number of people who no longer actively contribute to TDI, beginning with J.D. Wulfhorst and Nilsa Bosque-Pérez, who contributed significantly during the early years of this project. Other former TDI community members include Brian Crist, Ruth Dahlquist-Willard, Jesse Engebretson, Renée Hill, Justin Horn, Chris Looney, Ian O'Loughlin, Zachary Piso, Liela Rotschy, and Brianne Suldovsky.

Second, we have had the good fortune of collaborating with a number of very talented researchers and colleagues. We owe a great debt of gratitude to the following people, and anyone we inadvertently leave out: Janice Capel Anderson, Bert Baumgaertner, Philip Beesley, Melissa D. Begg, Eric Berling, Jan Boll, Evelyn Brister, NiCole Buchanan, Douglas Buhler, Ariane Cherbuliez, Lorenzo Ciannelli, Barbara Cosens, Bryan Cwik, Steven Daley-Laursen, Frank Davis, Tom Dietz, Kristie Dotson, Heather Douglas, Sandra Eckert, René Eschen, Matt Ferkany, James Foster, Robert Frodeman, Trish Glazebrook, Erik Goodman, Lissy Goralnik, Rob Gorbet, Paul Griffiths, Kathleen Halvorsen, Paul Hanson, Bill Hart-Davidson, Amanda J. Hessels, J. Britt Holbrook, Casey Rebecca Johnson, Jodi Johnson-Maynard, Samantha B. Joye, Machiel Keestra, Klaus Keller, D. Kerabatsos, John Kerr, Paul Kjellberg, Jonathan Kramer, Elaine L. Larson, J. Lebowski, Douglas Lind, Monica List, Christopher P. Long, Steven McGreevy, Matt McKeon, Barbara L. Miller, Doreen O'Connor-Gómez, Robert T. Pennock, Sara Pepper, Katie Plaisance, Aleta Quinn, Janet Rachlow, Emily Read, Urs Schaffner, Lynn Schnapp, Daniel Schoonmaker, Thomas P. Seager, Isis Settles, Elizabeth Simmons, George Smith, W. Sobchak, Ana Spalding, Daniel Steel, Lori Stinson, Dan Stokols, Dave Stone, Ajit Subramaniam, Jared Talley, Paul B. Thompson, Nancy Tuana, Andrew Turner, Sean Valles, Kathie Weathers, Kyle Powys Whyte, Holly Wichman, and Chris Williams.

Third, we thank Michigan State University (MSU), the University of Idaho (UI), and Boise State University (BSU) for financial and material support, and the National Science Foundation, from which we have received several awards (NSF SES-0823058, NSF SBE-1338614, NSF OIA-1954196, NSF OISE-1844794). Any opinions, findings, and conclusions or recommendations expressed in this book are those of the authors and do not necessarily reflect the views of these institutions or of the National Science Foundation.

Fourth, we thank the International Network for the Science of Team Science (INSciTS) and especially Holly Falk-Krzesinski, Steve Fiore, Gaetano Lotrecchiano, and Amanda Vogel. INSciTS has been an organizational home away from home for TDI for many years.

Finally, we thank other key partners for TDI: AgBioResearch (MSU), BEACON (MSU), Institute for Modeling Collaboration and Innovation (IMCI, UI), two UI Integrative Graduate Education and Research Traineeship (IGERT) projects,

ECOGIG, two UI NSF-EPSCoR projects, the Global Lake Ecological Observatory Network (GLEON), Institute for Bioinformatics and Evolutionary Studies (ibest, UI), Institute of Translational Health Sciences (ITHS, University of Washington), Knowledge Integration (University of Waterloo), Living Architecture Systems Group (LASG), Northwest Climate Science Center, Regional Approaches to Climate Change (REACCH), Risk and Uncertainty Quantification in Marine Science (Oregon State University), the Socio-Environmental Synthesis Center (SESYNC), the Sustainable Climate Risk Management Network (SCRiM), the Whittier Scholars Program (Whittier College), and Woody Weeds in East Africa.

Editors

Graham Hubbs is Chair of the Department of Politics and Philosophy at the University of Idaho. He has been a senior member of TDI since 2014. His research focuses on practical rationality, social ontology, and interdisciplinary communication. Within these broad areas, he specifically studies neo-Aristotelian accounts of human nature, the place of free speech and free press in democracy, and the ontology, history, and politics of money.

Michael O'Rourke is Professor of Philosophy and faculty in AgBioResearch and Environmental Science & Policy at Michigan State University. He is Director of the MSU Center for Interdisciplinarity (http://c4i.msu.edu/) and Director of TDI (http://tdi.msu.edu/). His research interests include epistemology, communication, and epistemic integration in collaborative, cross-disciplinary research, and linguistic communication between intelligent agents.

Steven Hecht Orzack is Senior Research Scientist and President of the Fresh Pond Research Institute (www.freshpond.org). His research interests include ecology, evolution, human health and disease, and the history and philosophy of biology.

Contributors

Stephen Crowley is a Professor of Philosophy at Boise State University. He has been part of TDI since 2007 and TDI's focus on both theoretical and applied approaches to facilitating cross-disciplinary research forms a large part of his research agenda. As far as he is concerned these interests flow naturally into his work on epistemology more generally, where he focuses on aspects of social epistemology using virtue theoretic perspectives and experimental methods.

Shannon Donovan is a founding faculty member and an Associate Professor of Sustainable Rural Systems at Eastern Oregon University. She enjoys working across fields of studies to address complex environmental problems and has been working as a cross-disciplinary practitioner for 17 years. Her areas of interest include environmental problem solving, bioregional planning, and interdisciplinary communication. She has a BS in wildlife management from the University of New Hampshire, an MS in parks, recreation, and tourism from West Virginia University and a PhD in environmental science from the University of Idaho.

Sanford D. Eigenbrode is Professor of Entomology and University Distinguished Professor at the University of Idaho. He conducts primary research on chemical ecology, landscape ecology and management of pests, beneficial insects, and insect vectors of plant pathogens affecting crops. The broader context of this work is food production as a social-ecological system, in which insects play a part. This has led to his involvement in and leadership of interdisciplinary projects addressing the sustainability of agricultural systems. As an outgrowth of his interdisciplinary endeavors, He works with philosophers to conduct research and provide consulting to improve communication within collaborative science and its applications.

Chad Gonnerman is an Associate Professor of Philosophy at the University of Southern Indiana. He has publications on the structural nature of concepts, the philosophical practice of deploying intuitions as evidence, the shape and content of folk epistemology and lay concepts of knowledge, the nature of interdisciplinary integration, and philosophy's ability to enhance cross-disciplinary research, among other topics.

Troy E. Hall is Professor and Head of the Department of Forest Ecosystems and Society at Oregon State University. She has degrees in anthropology and forest resources. She conducts research on public values and attitudes toward natural resource management issues and teaches courses on communication theory. She has worked with interdisciplinary teams of graduate students and researchers and has conducted research on the curricular needs of interdisciplinary teams. Her recent research has explored the effectiveness of different pedagogical approaches in enhancing learners' abilities to reason about complex socio-ecological problems.

Bethany K. Laursen is Principal Consultant and Owner of Laursen Evaluation & Design, LLC, where she empowers knowledge professionals to make sense of sustainability issues by using the right tool at the right time. Her research reveals how these tools work as aids to cross-disciplinary communication, cognition, and collaboration. Her strong understanding of cross-disciplinary process is rooted in her own formal training in seven disciplines, two inter-disciplines, and one trans-discipline, her five years as a K12 educator, and her seven years as an environmental evaluator. She is a research member of TDI.

Anna Malavisi is an Assistant Professor in the Department of Philosophy & Humanistic Studies at Western Connecticut State University. She has a PhD in philosophy from Michigan State University. Her dissertation, *Global Development and Its Discontents: Rethinking the Theory and Practice*, is a critical analysis of global development from an ethical and feminist epistemology perspective. Her interests include development and global ethics, feminist philosophy/epistemology, social and political thought, and the environment. She has a Master of Health and International Development and has worked for 16 years in Latin America in the NGO sector in areas of development practice and management.

Marisa A. Rinkus is a Postdoctoral Research Associate with the Center for Interdisciplinarity (C4I) at Michigan State University. Her work with C4I involves the design and facilitation of TDI workshops, as well as the study of structured dialogue in supporting cross-disciplinary communication and collaboration. Her areas of research include community engagement in wildlife/natural resource conservation, interdisciplinary collaboration, science of team science, gender, diversity and inclusion, and international development. She holds a PhD in fisheries and wildlife with a focus on human dimensions.

Brian Robinson is an Assistant Professor of Philosophy at Texas A&M University-Kingsville. Besides his work with TDI, his research focuses on moral psychology and philosophy of language. Part of the intersection of these two fields is in the moral psychology of humor and amusement. He is an inveterate punster, which he swears is more than just some antics. He lives in South Texas near the beach, which he enjoys when he can get there.

Stephanie E. Vasko is the Managing Director for the Center for Interdisciplinarity at Michigan State University. She holds a PhD in chemistry and nanotechnology (2012) and an MS in chemistry (2009). She builds community and collaborative capacity among academic, community, and blended academic/community research teams by developing and delivering philosophically informed team-based workshops as part of TDI. She also researches the application of machine learning to team science and artistic practice, and also conducts research around ceramics and mixed media sculptures. She was an American Association for the Advancement of Science Community Engagement Fellow.

1 Communication Failure and Cross-Disciplinary Research

Stephen Crowley
Michael O'Rourke

1.1 THE DIFFERENCE DILEMMA

In 2010, a snack food company decided that it wanted its products to be in the deli sections of supermarkets because deli sections are perceived to be the "cool" parts of supermarkets, with all that entails in regard to pricing and sales. The company hired a group of ethnographers to come up with compelling, research-based arguments for locating its products in the deli sections (Sunderland 2016). The ethnographers conducted interviews, did field observations, and ran focus groups. Their report focused on the meanings shoppers associated with delis and what these implied for their client's products. In Sunderland's words (p. 126):

> We discussed food as a form of cultural capital entailing trajectories of travel, discovery, adventure, or health, all of which we detailed with the experiences of our research participants. We talked about how delis fit practically, emotionally, and symbolically within the context of daily life. In a daylong presentation that included numerous illustrative edited video and audio excerpts of the ethnographic material, we then went on to focus more specifically on the dynamics of shopping deli areas versus interior aisles of stores and on the salient contexts for choosing specific delis. Finally, we focused on our client's product categories: how these products were semiotically constructed in delis and in grocery, and how these products were shopped and consumed.

Unfortunately, this way of thinking about delis and snack foods did not align with the client's way of thinking. For the client, questions about the value of locating snack foods in deli sections concerned basic economics—what needs were being met and how customers made their purchasing decisions. From the client's perspective, the key questions were (p. 127):

> What were deli "purchase drivers?" What "needs" were consumers fulfilling when they purchased items in the deli? How did they make their decisions (i.e., "What are consumers' decision trees?")? These were the questions he insisted needed answers; these, he maintained, were the questions of interest for both retailers and his corporation, and in our research presentation, we had not answered them.

The client was not prepared to consider the alternative framing of the material that the ethnographers presented, and so the final report prepared by the ethnographers was expressed only in the framework the client demanded. Consequently, the company's thinking about delis didn't change. Sunderland concludes that the company missed a chance to become a thought leader in this area.

We begin with this anecdote because it exemplifies what we call the "Difference Dilemma." One horn of this dilemma corresponds to research endeavors that emphasize the perspective of a single discipline. This emphasis has the virtue of streamlining the framing of the research problem, the language used to talk about it, and the methods used to respond to it, thereby avoiding the kind of failure described above. However, this single-discipline approach may reduce the ability of the research to address complex problems in ways that do justice to their complexity. The result can be a simplistic response that does more harm than good. The other horn is the cross-disciplinary one that embraces difference, choosing the prospect of making real progress in addressing a complex problem despite the very real potential of project failure. (In this book, we use "cross-disciplinary" generically to apply across the range of phenomena that could be described as multi-, inter-, or transdisciplinary; we will use a more specific term only when the specificity is required by the context.)

Once you are aware of the Difference Dilemma, you can find it in many contexts where people seek to understand complex phenomena (e.g., academic research projects, environmental conservation projects, or local planning efforts). In spite of the risks, task groups, institutions, funding agencies, and researchers increasingly choose the cross-disciplinary approach when trying to address complex problems (Van Noorden 2015). In fact, there are many projects that *require* different forms of expertise in order to make progress. In these cases, *difference* is critical—without it, these projects cannot do justice to their goals. However, the presence of these different forms of expertise increases the likelihood of failure. Although the ethnographers in the deli example met the company's requirements, their way of understanding the situation was so disconnected from the client's interests that it seemed like they had changed the subject.

The need to address complex problems continues to increase, and as a result interest in the Difference Dilemma has also risen over the past few decades. In particular, there has been significant discussion of issues at the heart of the Difference Dilemma. For instance, Campbell (2005), in discussing lessons she learned from her work as a social scientist collaborating with conservation biologists, describes both opportunities and challenges associated with an attempt to synthesize different visions of good research practice. For Campbell, the challenge is that (p. 575) "Social and natural scientists often approach conservation from different perspectives, both in terms of defining the problem and determining the appropriate approach to understanding it." Interdisciplinary collaboration can be very rewarding, since recognition of differences in how collaborators approach a problem can spur deeper reflection on one's own research approach and can also supply opportunities to create new connections and insights among disciplines. But it can also be very challenging, since different disciplinary perspectives are often so ingrained that it can be difficult for a researcher to identify and acknowledge them. If left unaddressed, such differences can undermine interdisciplinary collaboration, especially (p. 575) if they "arise unexpectedly and at an inopportune moment (e.g., when trying to finalize project outputs)."

Bracken and Oughton (2006) describe similar challenges when combining differing disciplinary expertise. They focus on how to build communication resources to facilitate successful interdisciplinary collaboration. For them, what is needed is (p. 373) "to understand the ways of thought and language of others." In an interdisciplinary project, differences are often manifested in the language used by collaborators. These differences are most obvious when we focus on disciplinary jargon that is mysterious to outsiders, but more insidiously they also reside under terms that seem to be held in common across disciplines. Bracken and Oughton discuss terms such as "dynamic" that are differently understood across disciplines. They address how different ways of thought are manifested in the language used within interdisciplinary projects and how shared understanding of these differences is key to taking advantage of multiple forms of expertise. Along with this potential advantage, however, comes the possibility of confusion that could undermine the project. Once again, in our terms, we find reflections of a choice forced by the Difference Dilemma. (For more on these issues, see Chapter 4.)

In summary, when we grapple with complex problems, we are likely to encounter the Difference Dilemma. When we do, we have two choices:

(1) Reject difference and adopt a uni-disciplinary approach to the problem, as the corporation did in the example above. This results in a partial understanding of the problem, at best.
(2) Embrace difference and adopt a cross-disciplinary approach to the problem. This results in a fuller understanding of the problem, when things go well, or confusion and misunderstanding, when things do not.

The desire to address and perhaps even solve a complex problem motivates selection of the cross-disciplinary horn of the dilemma. In doing so, one takes on risks that can play out as confusion, intractable disagreement, and deal-breaking divisiveness, all due to the disciplinary differences that one embraces in making this choice. This is a classic high risk/high reward vs. low risk/low reward dilemma, *unless we can change the odds* so that we manage differences to promote the likelihood of fuller understanding and reduce the likelihood of misunderstanding. This volume is devoted to deepening our understanding of the cross-disciplinary choice and suggesting ways of managing it so that we can successfully address the Difference Dilemma.

1.2 FAILURE LITERATURE

The deli example is a *failed* interdisciplinary project, in the sense that the researchers' preferred way of motivating the placement of products in the deli section was deemed a non-starter by the company that hired them. Many people who engage in collaborative cross-disciplinary work believe that cross-disciplinary efforts fail more frequently than disciplinary efforts (MacLeod 2018; Fam and O'Rourke 2020). This book is dedicated to helping cross-disciplinary projects avoid failure. It provides an account of the theoretical foundation and the practical aspects of an effort—the Toolbox Dialogue Initiative (TDI)—that both studies and facilitates cross-disciplinary research. As our opening vignette illustrates, cross-disciplinary projects can fail because partners in the project have difficulty communicating with one another. Through dialogue-based workshops designed to enhance collaborative communication, TDI supplies teams with resources they can use to address the challenge of communicating across differences and to avoid project failure.

As a first step to understanding how to prevent the failure of cross-disciplinary projects, it is helpful to know the various ways in which such projects can fail. First, there is *collapse*, which occurs when a project falls apart before it attains any objectives, often before they really get started (Norris et al. 2016). Collapse might occur because it takes too long to develop the mutual understanding required for successful interaction with your collaborators (Lélé and Norgaard 2005) or because of the difficulty of securing the mutual respect needed for full participation of all involved (Campbell 2005; Gardner 2013). These and other challenges can undermine the commitment and persistence required for cross-disciplinary success.

A cross-disciplinary project can also fail by not achieving project objectives. This can happen when objectives are altered while the project remains intact (e.g., O'Malley 2013). These adjustments to a project's objectives may be attributable to a number of factors, e.g., loss of key personnel, an overly ambitious schedule, or inability of the team to integrate their perspectives (Jakobsen et al. 2004). When it comes to meeting antecedently specified objectives, cross-disciplinary projects can be especially vexing, since it can be difficult to know in advance what the integrative capacity of the team will be or how the different perspectives will come together in integrative combination (Klein 2012; Salazar et al. 2012; Piso et al. 2016). In cases where failure to meet objectives does not lead to project failure, the desire to keep the team together, perhaps because of sunk costs and the need to get something out of one's effort, can override the motivation to leave this riskier mode of research practice and return to projects that are safer, such as disciplinary projects that present fewer obstacles.

Another key step in preventing failure is to learn from past failures. Learning from failure in cross-disciplinary research projects can be difficult. In many contexts, there is a premium attached to reporting success, which disincentivizes lingering over projects that don't work out. Projects may be encouraged to "fail fast" so that participants can move on to the next (hopefully) successful project (Babineaux and Krumboltz 2013). This does not leave time for writing up the reasons why a project failed, which can be a fraught exercise given the potential for damaging relationships with collaborators who may view the experience differently or who may not want a record of it.

However, as cross-disciplinary scholarship has grown in importance, published descriptions of failures in cross-disciplinary projects have increased. These are invaluable in the context of helping us learn from prior experiences, especially in regard to problems rooted in unappreciated differences among collaborators (Jakobsen et al. 2004; Della Chiesa et al. 2009). For example, there is a tendency on the part of collaborators to believe that they agree with one another—that they see their project in essentially the same way (Bracken and Oughton 2006). This positions them to explain away apparent difference (cf. Grice 1989), which can lead them to ignore important differences that deserve attention. What is missing in these cases is team reflexivity, understood as the capacity to support the recognition and consideration of differences among collaborators.

1.3 THE TOOLBOX DIALOGUE INITIATIVE

At this point you may be tempted to throw this book away and give up your plans to work with collaborators from other knowledge-making traditions. After all, we've described how such collaborations can be more challenging than disciplinary projects and, for this reason, more prone to failure. However, we encourage you to keep reading! In the pages that follow, we describe how and why the chances of success for cross-disciplinary collaboration can be increased.

We first ask you to consider the possibility that you and your potential collaborators think about the world more differently than you suspect. Furthermore, many of these differences in perspective may inform how you practice your research without you recognizing it. This can make those differences hard to negotiate. Some differences within cross-disciplinary teams are obvious, such as the disciplinary backgrounds of participants. Some are less obvious, such as the different ways that disciplinary background, understood to be the tacit operation of the core beliefs and values that one acquires in becoming a disciplinary expert, and personal experience shape the practice of team members in pursuit of team goals. Differences like these are particularly pressing when team goals are significantly open ended, as they often are on knowledge-making teams. If your goal is to leverage the differences among teammates, then an important first step to doing this systematically is to make these differences explicit and coordinate them.

TDI is devoted to understanding and facilitating this important first step. The centerpiece of this approach is the Toolbox workshop, a team activity organized around a structured dialogue that reveals team members' ways of knowing so that they become shared resources for use in subsequent cross-disciplinary activities. The philosophical concepts and methods used to structure the dialogue support the sort of introspection that can disclose foundational commitments such as the core beliefs and values that constitute disciplinary ways of knowing. Once disclosed, these commitments can then be coordinated in dialogue by collaborators. In this way, a key and well-noted obstacle to cross-disciplinary success can be revealed and avoided.

1.4 A BRIEF HISTORY OF THE TOOLBOX APPROACH

TDI originated in the "Ecosystem Management in Tropical and Temperate Regions: Integrating Education in Sustainable Production and Biodiversity Conservation" project at the University of Idaho (Bosque-Pérez et al. 2016). This project, funded by the US National Science Foundation (NSF), provided team-based interdisciplinary training for PhD students from agricultural and natural resource disciplines. (For more on how we use "interdisciplinary" here, see the discussion of cross-disciplinarity below.) The students were organized into interdisciplinary teams with a geographic focus, and each team worked to identify interdisciplinary research questions that could structure integrated research. Teammates worked toward producing dissertations that were coordinated around a common topic; each dissertation included one interdisciplinary chapter co-authored by all teammates. The emphasis on meaningful collaboration was reflected in the fact that successful completion of the PhD in the project was dependent on integrative success.

Three years into the project, the students were experiencing many of the problems of interdisciplinary collaboration, despite significant resources supporting their work. To help address this situation, the students asked for a seminar dedicated to philosophical issues in interdisciplinary research, thinking that philosophy could provide resources that would help them develop integrative research questions and find common ground for their collaborative efforts. This seminar was led by two of the authors contributing to this book, Eigenbrode and O'Rourke, and focused on developing something that could be used by interdisciplinary collaborators to improve their prospects for success. Over the course of the semester, the seminar participants developed the first version of the Toolbox approach, thinking of it as a tool that could help interdisciplinary teams communicate and integrate across disciplinary divides. A subset of seminar participants continued to work on the approach, introducing it into the academic literature with Eigenbrode et al. (2007). Subsequent NSF funding allowed TDI team members to conduct workshops for teams around the United States, focusing at first primarily on those doing environmental research. During the period of the NSF award, the Toolbox team conducted 76 Toolbox workshops and began to modify the approach, especially so that it could be extended to apply to health science, including community health and translational health.

Since those early days, TDI has grown into a research community, conducted many more workshops, and expanded its range and influence. As of this writing, TDI has conducted over 300 workshops with research teams, research communities, undergraduate and graduate courses, training groups, and ad hoc groups of participants. These workshops have taken place in 23 states and 16 countries. Most have involved scientific teams, courses, or communities, although recently TDI has conducted more workshops that involve community partners either working alongside researchers or conducting projects on their own.

In all of these cases, the prerequisite for a Toolbox workshop is that there are differences in perspective represented in the group that are relevant to how the group functions (O'Rourke and Crowley 2013). These differences are rooted in the differing values and beliefs, often implicit, that are acquired as part of acquiring the perspective itself, e.g., when a disciplinary expert is in graduate school and working toward becoming an expert. The goal of the workshops is to give the participants a chance to reflect collectively on their differences and their similarities so that they are able to observe their common problem from each other's perspective (Looney et al. 2014). (For more on the history of TDI, see Chapter 3.)

1.5   CORE IDEAS

The Toolbox approach creates opportunities for collaborators to reflect on their practice, share those reflections with one another, and expand their mutual understanding. These reflections take place in a workshop that is structured by a set of philosophically grounded prompts, which we refer to as a "Toolbox Instrument." For example, in the original "Scientific Research Toolbox Instrument," used in 98 of the more than 300 workshops TDI has run, there are prompts that address methodology, such as "Scientific research (applied or basic) must be hypothesis driven," and prompts that address confirmation, such as "Validation of evidence requires replication" (see Appendix A and Looney et al. 2014). Examination of their varied responses to these prompts gives a group an opportunity to work out the pattern of agreement and disagreement associated with each of these foundational commitments. Awareness of this pattern enhances mutual understanding within the team, allowing it to improve communication and improve prospects for project integration. While there is variation between particular workshops as we fit the approach to important aspects of the local context, there are a number of core themes that remain constant. The core themes that animate the Toolbox approach are *cross-disciplinarity*, *dialogue*, *philosophy*, *mutual understanding*, *communication*, and *integration*. In this section we sketch each of these ideas, which are developed in greater depth in subsequent chapters.

1.5.1 CROSS-DISCIPLINARITY

A good way to make sense of cross-disciplinarity is to consider disciplines as communities. On this way of thinking, to be a member of discipline X is to share a way of life with the other members of that discipline, centered on a set of intellectual challenges, practices, and methods, which come together to mark off a distinctive disciplinary "worldview" (O'Rourke and Crowley 2013; Repko et al. 2017; see also Chapter 5). We tend to think of disciplines as separate from one another, which encourages talk of "boundary crossing" when we want to acknowledge a person or project that draws on the resources of more than one discipline (Nicolescu 2002). Cross-disciplinarity crosses the boundaries of disciplines in order to combine aspects of their different worldviews.

Cross-disciplinary work can be accomplished by both individuals and groups, and indeed groups of groups, networks, and so on. An engineer thinking about the cultural impact of their work is engaged in cross-disciplinarity of the same sort that you could find on a team of engineers and sociologists. We focus on the team context because the unique challenges associated with doing *collaborative*, cross-disciplinary work are the focus of our research.

As we noted above, we use "cross-disciplinarity" as a generic term that applies to multidisciplinarity, interdisciplinarity, and transdisciplinarity. These terms and others like them (O'Rourke et al. 2019) correspond to different ways in which disciplinary inputs can be combined. If the inputs alter little (e.g., in a report that includes sections on culture, economics, and hydrology), then the work is multidisciplinary. If the ideas alter more significantly (e.g., cultural, economic, and hydrological factors all included in a single model), the work is interdisciplinary. Finally, if the ideas alter to the point of novelty (e.g., the cultural, economic, and hydrological perspectives give rise to a new field), the work is transdisciplinary. For more on this point, see the discussion of *integration* below and in Chapter 5. ("Transdisciplinarity" is also used to indicate the involvement in research of non-academic collaborators. See Klein 2017.)

Our unit of analysis, the notion of *discipline*, is often used to identify administrative structures, such as academic units in a university like the Department of Philosophy or the Department of Chemistry. However, not all academic units contain people who share worldviews—it is common to find departments that amalgamate several different disciplines. The notion of *discipline* is also used to pick out groups of people connected with professional organizations and associated community structures, such as conferences and journals, that advance the work of those who share their worldview. Here the administrative structure is the professional society, which is typically large in membership and international in distribution.

In this last sense, *discipline* has much in common with the notion of *epistemic community*, which is a group of people who share an interest in a certain set of intellectual challenges and in ways of working to meet those challenges (Haas 1992). Not all epistemic communities are disciplines, however; members of a particular lab may form an epistemic community, for example, but that community is not sufficiently large or stable across time to count as a discipline. Disciplines, by contrast, are epistemic communities that are stable enough to propagate (Turner 2000) and significant enough to give rise to academic units or professional societies.

1.5.2 DIALOGUE

TDI has been committed to dialogue from the beginning. The idea behind the Toolbox approach is that *structured* dialogue—i.e., talking together in an engaged way about the right things—can enable collaborators to collectively discover aspects of their work together that might have been previously unappreciated. We think of the Toolbox approach as a particular dialogue method (McDonald et al. 2009): it is a way of organizing conversation to support the creation of understanding in pursuit of project integration.

Dialogue can be contrasted with *discussion*, which is less engaged and interactive than dialogue. Both theoretical and empirical research emphasize that to call something *dialogue* is to emphasize that interlocutors are committed to cooperating with one another, building a type of communicative reciprocity that is reflected in charitable speaking and listening (Clark 1996; Traxler 2012). Dialogue partners listen carefully, motivated by a commitment to understanding that generates common ground and enables dynamic joint construal (Bakhtin 1981; Clark 1996). Whereas discussion can be superficial, dialogue can serve as a mechanism for generating deep insight through conversational effort.

The Toolbox workshop promotes this kind of conversational engagement. By focusing conversation on issues of common concern and creating an explicit environment for group reflection, the Toolbox workshop creates a context in which people are motivated to engage in deep listening, to clarify their similarities and differences, and to negotiate collective interpretations of their common project. Of course, dialogue is not always achieved—it can be difficult, and you can't guarantee that everyone will be up for it. In those workshops where it is achieved, however, gains are often made that enhance a team's collaborative and communicative capacity (see Chapter 7).

1.5.3 PHILOSOPHY

As we have noted, the practice of philosophy figures centrally in the Toolbox dialogue method. Specifically, philosophical concepts and methods are utilized in creating the dialogue-based workshop experience that is the central mechanism of capacity enhancement. Given this, a word or two about what we take *philosophy* to be is in order.

A Google® search turns up a variety of definitions of philosophy. Many overlap with this one: "The study of the fundamental nature of knowledge, reality, and existence, especially when considered as an academic discipline." So understood, it is not clear what philosophy provides researchers from different disciplines who try to coordinate their disciplinary worldviews in a constructive way.

Google also supplies definitions of philosophy similar to this one: "The study of the theoretical basis of a particular branch of knowledge or experience." Conceived in this way, philosophy is the study of what sorts of things a particular branch of knowledge or experience takes seriously, how people learn about those things, what the goals of the practice that supports the branch of knowledge or experience might be, and what it is to carry out the practice well. Philosophy of science is an instructive example of this way of understanding philosophy—it asks questions like, "What makes science *science*?" and "What is it to do science well?"

This way of understanding philosophy helps clarify its value to cross-disciplinary research. (For more on the value of philosophy to cross-disciplinary research, see Chapter 6.) Think of a disciplinary practice as analogous to a game with rules we do not understand. Philosophy aims to work out those rules from observation of and reflection on the disciplinary practice. These would include *constitutive* rules that determine what moves are allowable and thereby *create* the practice, and *regulative* rules that "regulate a pre-existing activity" (Searle 1969, p. 34) and so can be used to distinguish better forms of disciplinary practice from worse. Take chess for example. It is a constitutive rule that you can't expose your king and put yourself in checkmate. Such a move is illegal, in that it is part of what makes chess *chess* that you can't do this. In contrast, a regulative rule is that it is usually a bad idea to expose your queen and allow it to be taken; that is, such a move is legal but rarely part of successful chess-playing.

An analogy may help us understand the value of philosophy for cross-disciplinary practice. Imagine that different disciplines are like different styles of dancing. Researchers are expert dancers of their particular disciplinary style. The challenge of cross-disciplinary work is to get expert dancers of different styles to dance well together. Disciplinary experts dance without thinking about the hows and whys of their dance. But it is just these aspects—the hows and whys—that are critical when you are attempting to dance with someone who dances a different style. Reflection on these aspects of dance can illuminate how to coordinate and choreograph the collective dance. In the cross-disciplinary context, philosophy is helpful because the hows and whys of various practices are exactly what philosophical evaluation of these practices aims to articulate. So philosophy can provide resources that allow experts to identify and articulate the basis of their expertise, which enables other experts to recognize how they might coordinate their practices and produce a graceful collaboration.

1.5.4 MUTUAL UNDERSTANDING

Philosophically structured dialogue is valuable for a cross-disciplinary team because it can enhance mutual understanding, which is an achievement that is beneficial for team communication and project integration (see Chapter 10). As a venue for structured reflection and active perspective seeking, a Toolbox workshop encourages collaborators to tell one another what they think about fundamental aspects of their work together, which can reveal points of view and commitments that may be news to the group. This results in mutual understanding of how each team member thinks and acts as a researcher or practitioner in the context of their common project.

Mutual understanding conceived in this way requires that knowledge about each collaborator be held in common by the team, but it also requires that the collaborators are aware that they share this knowledge in common. This mutual understanding is a form of what is known in the epistemology literature as *common knowledge* (Lewis 1969; Schiffer 1972).

Mutual understanding is also related to *common ground*, a theoretical construct that has roots in literature that concerns communication and conversational interpretation (Stalnaker 1978; Lewis 1979; Clark 1996). This construct has been deployed in the interdisciplinary theory literature to express "the basis for collaborative communication across disciplines and integration of their conflicting concepts or theories" (Repko 2012, p. 322). In this literature, common ground is modeled as a cognitive achievement that consists of common knowledge of disciplinary assumptions and supports the coordination and integration of disciplinary insights. Work to achieve common ground, so understood, describes the kind of mutual understanding we strive to achieve in Toolbox workshops.

It is important to note here that mutual understanding does not entail agreement. Workshop dialogue is intended to illuminate similarities and differences among collaborators so that they can be coordinated and managed. It is *not* intended to wash differences away in favor of a uniform consensus. After all, the point of cross-disciplinary research is to leverage differences among disciplinary contributors so as to increase the complexity and adequacy of the research response.

The Toolbox approach generates conditions conducive for gains in mutual understanding by creating a teaching-and-learning environment featuring conversation about the assumptions, values, and priorities that frame each collaborator's contribution to the project. These conversations enable collaborators to teach each other how they approach their common effort, which in turn supports project-wide learning about how their teammates function. Individual reflexivity informs the teaching, and collective reflexivity is enhanced by the learning that takes place in these workshops. The result is that team members are in a better position to view their project from each other's perspectives (Looney et al. 2014).

1.5.5 COMMUNICATION

The goal of Toolbox workshops is to improve communication within cross-disciplinary teams (Looney et al. 2014). By learning more about each other, collaborators develop a better sense for when a teammate might have a view on an issue or when a teammate is well-positioned to address a question. This heightened sensitivity to one's collaborators reduces the potential for misunderstanding and increases the potential for more efficient distribution of communicative labor and knowledge exchange. More effective communication can create a greater degree of team cohesion and collective communication competence, and support the team in pursuing its integrative goals (Casey-Campbell and Martens 2009; Thompson 2009; Klein 2013).

It is easy, but mistaken, to think of communication simply in terms of moving bits of information around. A richer notion of communication closely connected with collaboration is critical to thinking clearly about its role in Toolbox workshops. Collaboration involves intentional interaction in pursuit of a goal, and what differentiates *inter*action from parallel action is the mutual dependence of behavior; that is, two people *collaborate* when each person's actions are partially influenced by the actions of the other in a way that creates mutual interdependence (O'Rourke et al. 2014). Successful collaborations are typically not choreographed in advance, and so depend on collaborators observing each other's actions and gathering information that influences subsequent actions. Collaborators expect that their teammates will work with them and not against them, and so they act in ways that will aid the joint pursuit of their collaborative goals. This type of action can be understood as *communication*, which is usefully conceived as the co-creation of meaning in pursuit of a goal (Hall and O'Rourke 2014).

Characterizing communication as the fabric of collaborative interaction highlights two of its central aspects: the *sociopragmatic* and the *informational* (see Chapter 5; see also Watzlawick et al. 1967; Fisher 1979). The sociopragmatic aspect pertains to the relationships among communication partners that form the trust needed for mutual observation, while the informational aspect pertains to the information that is exchanged and upon which behavioral coordination depends. Co-creation requires at least a temporary relationship sufficient to enable information transactions and the joint development of an interpretation. If this is done in pursuit of a goal, the relationship begins to feel like a partnership that is built on trust.

Communication can be especially problematic when collaborators have substantially different worldviews. This is typically the case in cross-disciplinary contexts, where communication style and technical vernacular can vary dramatically across disciplines. For an example on the sociopragmatic side, consider a philosopher who wishes to create an adversarial relationship with collaborators, expecting progress to be made through critical argument. An engineer, by contrast, might desire a more harmonious relationship with communication that supports joint design and construction. On the informational side, differences in vernacular constitute a well-documented problem for collaborators seeking to work across disciplines (Bracken and Oughton 2006; Donovan et al. 2015). Toolbox workshops are intended to address challenges along both of these fronts.

1.5.6 INTEGRATION

Cross-disciplinary projects bring insights from different disciplines together in one place in order to address a common research question or achieve a common goal. Cross-disciplinary projects are thus integrative. This includes multidisciplinary projects, which as noted above are subsumed under cross-disciplinarity. Although there may be little or no interdependence among disciplinary contributions in a multidisciplinary project, the fact that they all are part of a single project indicates at least a minimal level of integration (cf. Kline 1995). We return to this in Chapter 5.

As we have suggested, the increase in mutual understanding among collaborators that Toolbox workshops strive to produce is expected to enhance the ability of those collaborators to integrate their disciplinary insights. It is clear that without mutual understanding—that is, without appreciation for what one's collaborators can contribute to the collective effort—it is very difficult to identify, compare, and combine those contributions. Mutual understanding, or common ground, increases the ability of collaborators to anticipate what each other says and does, enabling the planning of joint action and the identification of opportunities to combine their expertise.

We view integration as a process that produces an output by putting inputs into certain kinds of relations (O'Rourke et al. 2016). The key is the *integrative relation*, which can be contrasted with disintegrative or preservative relations. Integrative relations exhibited by integrated outputs typically place inputs (or intermediate products) into mutual dependence and reduce the number of outputs relative to inputs. Examples of integrative relations produced in the integrative process include *serialization*, *linking*, and *absorption*. The integration of two inputs can involve bringing them together, but it can also involve just a change in perspective that enables one to see the inputs as integrated. So understood, integrative activity takes place at many moments and at the individual, interpersonal, and project levels in a cross-disciplinary project. Integration of the sort that marks the project as *cross-*disciplinary requires putting disciplinary inputs into integrative relations that are evident in project outputs.

Integration so conceived occurs in degrees. One way to think about this is in terms of *mutual dependence*, with processes that are more integrative yielding greater mutual dependence among inputs. Another way is from the perspective of the output, with more integration corresponding to a greater degree of difficulty in recovering the inputs from the outputs. How much integration will be required will vary within a project and across projects. The Toolbox approach is designed to support focused interventions that contribute to the integrative goals of a project, whatever those might be. The coordination of perspectives that can be gained in Toolbox workshops, along with the opportunity to adopt the perspectives of one's collaborators, increases the ability of collaborators to determine where fruitful combinations of disciplinary insights can be made.

1.6 AN OVERVIEW OF THIS BOOK

TDI advocates and supports a philosophical, dialogue-based approach to addressing the challenges of collaborative cross-disciplinary research. This book is our most comprehensive development of the Toolbox approach to date, one that locates it in its theoretical context and develops it in a way that highlights what it can achieve. It is intended for collaborators on a cross-disciplinary team that seek to improve their collaboration, and also anyone interested in cross-disciplinary theory and practice, engaged philosophy, and the philosophy of communication.

The book can be thought of as consisting of three sections. The first consists of this chapter along with Chapter 2, "How It Works: The Toolbox Dialogue Method in Practice," which builds on the theoretical developments in the present chapter and lays out the practical details of addressing cross-disciplinary communication challenges, and Chapter 3, "A Narrative History of the Toolbox Dialogue Initiative," which offers an account of TDI history in order to make clear the interplay of the theoretical and practical aspects of our response with the challenges of cross-disciplinary research.

The second section contains four chapters. Chapter 4, "The Landscape of Challenges for Cross-Disciplinary Activity," presents the context within which the Toolbox dialogue method operates by supplying a general survey of the challenges that confront cross-disciplinary work. This survey suggests a response to these challenges, namely, organizing the differing worldviews at work in collaborative, cross-disciplinary research so that they "harmonize" with one another. We think of the requisite level of "harmony" as corresponding to the extent to which disciplinary inputs are integrated, and this can only be achieved if collaborators communicate their worldviews effectively. Chapter 5, "Communication and Integration in Cross-Disciplinary Activity," discusses work that is involved in integration and communication. Chapter 6, "The Power of Philosophy," highlights the potential of philosophy to aid the sort of communication and integration needed for efficient and effective cross-disciplinary work. Chapter 7, "The Power of Dialogue," emphasizes the potential of dialogue to serve as a mechanism that develops the mutual understanding that can support cross-disciplinary integration.

The Toolbox workshop is the focus of the third section. Chapter 8, "Best Practices for Planning and Running a Toolbox Workshop," provides details about "state-of-the-art" Toolbox workshops, focusing on how TDI currently plans and delivers these for a wide range of partners. Chapters 9 and 10, "Enhancing Cross-Disciplinary Science through Philosophical Dialogue: Evidence of Improved Group Metacognition for Effective Collaboration" and "Qualitative Analyses of the Effectiveness of Toolbox Dialogues," take a closer look at the data we've collected in running these workshops, outlining quantitative and qualitative evidence for the influence of the workshop approach on cross-disciplinary research teams. In Chapter 11, "Future Directions for the Toolbox Dialogue Initiative," we discuss future directions for the Toolbox approach. Finally, Chapter 12 presents a collection of case studies that bring out details of the Toolbox dialogue method in action. These accounts, along with the associated material in Appendices A and B, will give the reader a close and detailed perspective of what the Toolbox approach amounts to and what it achieves.

In this book, the authors tell the story of TDI in a variety of ways so as to allow the reader to focus on those topics and perspectives they expect to be most useful. Readers interested in the theoretical background of communication interventions in a cross-disciplinary context can focus on the second section, while those interested in the details of a specific tool for building communicative capacity in cross-disciplinary teams can focus on the first and third sections. Philosophers who are curious about how to engage with representatives of different disciplines can profit from the first and second sections. Of course, we invite readers to explore the entire book!

1.7 ACKNOWLEDGMENTS

We would like to thank Graham Hubbs and Steven Orzack for their editing assistance and all of our co-authors in this volume for their contributions, since without them there would be nothing to introduce in this opening chapter. O'Rourke's work on this chapter was supported by the USDA National Institute of Food and Agriculture, Hatch Project 1016959.

1.8 LITERATURE CITED

Babineaux, R., and J. D. Krumboltz. 2013: Fail Fast, Fail Often: How Losing Can Help You Win. Jeremy P. Tarcher/Perigee, New York.

Bakhtin, M. M. 1981: Discourse in the novel. Pp. 259–422 *in* The Dialogic Imagination. University of Texas Press, Austin.

Bosque-Pérez, N. A., P. Z. Klos, J. E. Force, L. P. Waits, K. Cleary, P. Rhoades, S. M. Galbraith, A. L. B. Brymer, M. O'Rourke, S. D. Eigenbrode, B. Finegan, J. D. Wulfhorst, N. Sibelet, and J. D. Holbrook. 2016: A pedagogical model for team-based, problem-focused interdisciplinary doctoral education. BioScience 66:477–488.

Bracken, L. J., and E. A. Oughton. 2006: "What do you mean?" The importance of language in developing interdisciplinary research. Transactions of the Institute of British Geographers 31:371–382.

Campbell, L. M. 2005: Overcoming obstacles to interdisciplinary research. Conservation Biology 19:574–577.

Casey-Campbell, M., and M. L. Martens. 2009: Sticking it all together: A critical assessment of the group cohesion–performance literature. International Journal of Management Reviews 11:223–246.

Clark, H. H. 1996: Using Language. Cambridge University Press, Cambridge.

Della Chiesa, B., V. Christoph, and C. Hinton. 2009: How many brains does it take to build a new light: Knowledge management challenges of a transdisciplinary project. Mind, Brain, and Education 3:17–26.

Donovan, S. M., M. O'Rourke, and C. Looney. 2015: Your hypothesis or mine? Terminological and conceptual variation across disciplines. SAGE Open 5:1–13.

Eigenbrode, S. D., M. O'Rourke, J. D. Wulfhorst, D. M. Althoff, C. S. Goldberg, K. Merrill, W. Morse, M. Nielsen-Pincus, J. Stephens, L. Winowiecki, and N. A. Bosque-Pérez. 2007: Employing philosophical dialogue in collaborative science. BioScience 57:55–64.

Fam, D., and M. O'Rourke. 2020: Interdisciplinary and Transdisciplinary Failures as Lessons Learned: A Cautionary Tale. Routledge, Abingdon, UK.

Fisher, B. A. 1979: Content and relationship dimensions of communication in decision-making groups. Communication Quarterly 27:3–11.

Gardner, S. K. 2013: Paradigmatic differences, power, and status: A qualitative investigation of faculty in one interdisciplinary research collaboration on sustainability science. Sustainability Science 8:241–252.

Grice, H. P. 1989: Studies in the Way of Words. Harvard University Press, Cambridge, MA.

Haas, P. M. 1992: Introduction: Epistemic communities and international policy coordination. International Organization 46:1–35.

Hall, T. E., and M. O'Rourke. 2014: Responding to communication challenges in transdisciplinary sustainability science. Pp. 119–139 *in* K. Huutoniemi and P. Tapio, eds. Heuristics for Transdisciplinary Sustainability Studies: Solution-Oriented Approaches to Complex Problems. Routledge, Oxford.

Jakobsen, C. H., T. Hels, and W. J. McLaughlin. 2004: Barriers and facilitators to integration among scientists in transdisciplinary landscape analyses: A cross-country comparison. Forest Policy and Economics 6:15–31.

Klein, J. T. 2012: Research integration: A comparative knowledge base. Pp. 283–298 *in* A. F. Repko, W. H. Newell, and R. Szostak, eds. Interdisciplinary Research: Case Studies of Integrative Understandings of Complex Problems. SAGE, Thousand Oaks, CA.

———. 2013: Communication and collaboration in interdisciplinary research. Pp. 11–30 *in* M. O'Rourke, S. J. Crowley, S. D. Eigenbrode, and J. D. Wulfhorst, eds. Enhancing Communication & Collaboration in Crossdisciplinary Research. SAGE, Thousand Oaks, CA.

———. 2017: Typologies of interdisciplinarity: The boundary work of definition. Pp. 21–34 *in* R. Frodeman, J. T. Klein, and R. C. S. Pacheco, eds. The Oxford Handbook of Interdisciplinarity, 2nd edition. Oxford University Press, Oxford.

Kline, S. J. 1995: Conceptual Foundations for Multidisciplinary Thinking. Stanford University Press, Stanford, CA.

Lélé, S., and R. B. Norgaard. 2005: Practicing interdisciplinarity. BioScience 55:967–975.

Lewis, D. K. 1969: Convention. A Philosophical Study. Harvard University Press, Cambridge, MA.

———. 1979: Scorekeeping in a language game. Journal of Philosophical Logic 8:339–359.

Looney, C., S. Donovan, M. O'Rourke, S. J. Crowley, S. D. Eigenbrode, L. Rotschy, N. A. Bosque-Pérez, and J. D. Wulfhorst. 2014: Seeing through the eyes of collaborators: Using Toolbox workshops to enhance cross-disciplinary communication. Pp. 220–243 *in* M. O'Rourke, S. J. Crowley, S. D. Eigenbrode, and J. D. Wulfhorst, eds. Enhancing Communication and Collaboration in Interdisciplinary Research. SAGE, Thousand Oaks, CA.

MacLeod, M. 2018: What makes interdisciplinarity difficult? Some consequences of domain specificity in interdisciplinary practice. Synthese 195:697–720.

McDonald, D., P. Deane, and G. Bammer. 2009: Research Integration Using Dialogue Methods. Australian National University Press, Canberra.

Nicolescu, B. 2002: Manifesto of Transdisciplinarity. SUNY Press, Albany, NY.

Norris, P. E., M. O'Rourke, A. S. Mayer, and K. E. Halvorsen. 2016: Managing the wicked problem of transdisciplinary team formation in socio-ecological systems. Landscape and Urban Planning 154:115–122.

O'Malley, M. A. 2013: When integration fails: Prokaryote phylogeny and the tree of life. Studies in History and Philosophy of Science Part C: Studies in History and Philosophy of Biological and Biomedical Sciences 44:551–562.

O'Rourke, M., and S. J. Crowley. 2013: Philosophical intervention and cross-disciplinary science: The story of the Toolbox Project. Synthese 190:1937–1954.

O'Rourke, M., S. J. Crowley, and C. Gonnerman. 2016: On the nature of cross-disciplinary integration: A philosophical framework. Studies in History and Philosophy of Science Part C: Studies in History and Philosophy of Biological and Biomedical Sciences 56:62–70.

O'Rourke, M., S. J. Crowley, S. Eigenbrode, and J. D. Wulfhorst. 2014: Introduction. Pp. 1–10 *in* M. O'Rourke, S. J. Crowley, S. D. Eigenbrode, and J. D. Wulfhorst, eds. Enhancing Communication and Collaboration in Interdisciplinary Research. SAGE, Thousand Oaks, CA.

O'Rourke, M., S. J. Crowley, B. K. Laursen, B. Robinson, and S. E. Vasko. 2019: Disciplinary diversity in teams, integrative approaches from unidisciplinarity to transdisciplinarity. Pp. 21–46 *in* K. L. Hall, A. L. Vogel, and R. T. Croyle, eds. Advancing Social and Behavioral Health Research through Cross-Disciplinary Team Science: Principles for Success. Springer, Berlin.

Piso, Z., M. O'Rourke, and K. C. Weathers. 2016: Out of the fog: Catalyzing integrative capacity in interdisciplinary research. Studies in History and Philosophy of Science Part A 56:84–94.

Repko, A. F. 2012: Interdisciplinary Research: Process and Theory. SAGE, Thousand Oaks, CA.

Repko, A. F., R. Szostak, and M. P. Buchberger. 2017: Introduction to Interdisciplinary Studies. SAGE, Thousand Oaks, CA.

Salazar, M. R., T. K. Lant, S. M. Fiore, and E. Salas. 2012: Facilitating innovation in diverse science teams through integrative capacity. Small Group Research 43:527–558.

Schiffer, S. R. 1972: Meaning. Oxford University Press, Oxford.

Searle, J. R. 1969: Speech Acts: An essay in the Philosophy of Language. Cambridge University Press, Cambridge.

Stalnaker, R. 1978: Assertion. Syntax and Semantics 9:315–332.

Sunderland, P. 2016: The cry for more theory. Pp. 122–135 *in* Advancing Ethnography in Corporate Environments: Challenges and Emerging Opportunities. Routledge, New York.

Thompson, J. L. 2009: Building collective communication competence in interdisciplinary research teams. Journal of Applied Communication Research 37:278–297.

Traxler, M. J. 2012: Introduction to Psycholinguistics: Understanding Language Science. John Wiley & Sons, Oxford.

Turner, S. 2000: What are disciplines? And how is interdisciplinarity different? Pp. 46–65 *in* P. Weingart and N. Stehr, eds. Practising Interdisciplinarity. University of Toronto Press, Toronto.

Van Noorden, R. 2015: Interdisciplinary research by the numbers. Nature 525:306–307.

Watzlawick, P., J. B. Bavelas, and D. D. Jackson. 1967: Pragmatics of Human Communication. WW Norton & Company, New York.

2 How It Works
The Toolbox Dialogue Method in Practice

Michael O'Rourke
Stephen Crowley

2.1 INTRODUCTION

The primary aim of this chapter is to help the reader understand the Toolbox experience. We describe the value of participating in a Toolbox workshop and what it is like to be involved in such a workshop. We identify the general features of the Toolbox *approach* and the motivations behind it, and we describe the current Toolbox *workshop* available from TDI. We also explain the key ideas behind the workshop as well as the primary aspects of the experience, viz., the Toolbox instrument, the dialogue, and the co-creation activity. Further detail about running Toolbox workshops and best practices are the focus of Chapter 8.

A second aim of this chapter is to describe the research orientation of TDI and how it relates to our workshops. We collect data from the teams that participate in workshops and use these data to investigate the impact of the workshop on its participants, the conditions that lead to successful Toolbox workshops, and the dynamics of cross-disciplinary communication and collaboration more generally. These investigations are part of a research feedback loop: our research generates insights that inform the Toolbox workshops, which in turn generate data that support more research into the intervention itself and cross-disciplinary work in general (O'Rourke and Crowley 2013; Crowley et al. 2016). See Figure 2.1. This feedback loop drives TDI's evidence-based research and outreach, ensuring that the work we do is engaged with the cross-disciplinary literature and with our partners' needs.

2.2 GENERAL CONSIDERATIONS

The Toolbox approach centers on dialogue-based workshops designed to enhance communication and collaboration in complex, cross-disciplinary projects through the enhancement of mutual understanding. Unacknowledged differences among collaborators concerning the implicit beliefs and values that animate their project contributions can lead to mistaken assumptions, unreasonable expectations, and disagreements. These differences can be insidious, hiding underneath the committed desire of collaborators to make their collaboration work, which can incline them to see agreement where there really isn't any. By making implicit commitments explicit, the team can reduce the potential for unpleasant surprises and increase its ability to

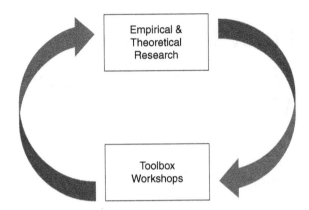

FIGURE 2.1 The research feedback loop that is at the heart of TDI as a research and outreach concern.

integrate its knowledge (Okhuysen and Eisenhardt 2002). An intervention such as the Toolbox approach can be the vehicle for doing this.

The main Toolbox mechanism for making the implicit explicit is a structured dialogue that focuses on a circumscribed range of topics of special interest to the collaborators (O'Rourke and Crowley 2013). The specific topics are chosen because of their potential to illuminate unacknowledged differences, as well as unacknowledged similarities among implicit beliefs and values, thereby allowing collaborators to coordinate their beliefs and values as well as negotiate them if necessary. As we will discuss in Chapter 6, philosophical concepts and methods play a role in structuring the dialogues. Philosophical concepts help map the common ground collaborators share as members of the same project. Philosophical methods support close analysis of disciplinary and professional commitments and facilitate movement toward common ground they share as researchers. Philosophy serves here as a lens that quickly and accurately identifies issues that require discussion, thereby enabling more effective communication and greater project integration.

In harnessing philosophy in this way, we strive to create dialogue in Toolbox workshops that deepens the appreciation collaborators have for each other's research and that facilitates perspective-taking without forcing them to grapple with potentially contentious project details. There are two motivations behind selecting this level of conversation. First, we want to structure the dialogue in a way that abstracts from the project itself, so as not to reproduce the misunderstandings that could or do occur, but not be so abstract that participants lose contact with their research and practice worldviews. We focus these dialogues on topics that support development of epistemic common ground and mutual awareness about the beliefs and values that shape how one operates as a researcher or practitioner. Second, we keep workshop participants close enough to their project that they regard the dialogue as relevant to their work. Otherwise, the workshop risks veering away from the epistemic issues that matter. This balance of abstraction and relevance motivates our selection of this intermediate level of conversation.

Philosophical analysis guides development of prompts that help Toolbox dialogue participants think about aspects of their research practice that may not be readily open to view (Eigenbrode et al. 2007). In the dialogue, these participants—who typically hail from different disciplines or professions—articulate, share, and coordinate their views on the prompts, enhancing mutual understanding by making what was an implicit part of the process of team deliberation explicit. This can remove obstacles to communicative efficacy and integrative success. Dialogue as a mechanism works by enhancing individual and team reflexivity, i.e., the ability to identify and consider aspects of one's own research worldview (Lewis et al. 2007), and perspective-seeking (or perspective-taking) behavior (see Chapter 10), i.e., the willingness and ability to adopt other points of view (Hoever et al. 2012). These enhanced abilities are attributes of effective teams that can "positively influence interdisciplinary team performance" (Salazar et al. 2019, p. 316).

A limitation of dialogue-based approaches is that the insights generated may be lost if they are not quickly converted into a form more lasting and stable than conversation. Further, it is important to ensure that workshop participants have something tangible to show for their efforts. When time is precious and process-focused efforts like Toolbox workshops take time away from experiments and manuscripts, it is important that teammates have more to show for their participation in the workshop than memories of a conversation, no matter how stimulating that might have been. To address these concerns, Toolbox workshops include co-creation activities following the initial dialogue sessions to ensure that dialogue insights are captured in stable and team-relevant artifacts.

Finally, Toolbox workshops are intended primarily for *teams*, although many workshops have been run for *groups* that do not consist of project collaborators. Such groups don't interact in ways that are functionally interdependent, and they don't necessarily identify as a collective outside of the workshop (cf. Fiore 2008). Although the Toolbox approach has value for such less cohesive collectives, it is designed primarily for more cohesive teams.

2.3 SO YOU WANT A TOOLBOX WORKSHOP? DETAILS OF THE WORKSHOP EXPERIENCE

We now discuss the elements of a Toolbox workshop. For illustrations, we provide details from a 2015 Toolbox workshop with the Research Institute for Humanity and Nature (RIHN) in Kyoto, Japan.

2.3.1 INITIAL CONTACT AND IDENTIFICATION OF WORKSHOP OBJECTIVES

When a team contacts TDI and requests a workshop, the initial conversations focus on identifying what work a Toolbox workshop might do for them. The following questions are posed to prospective workshop teams: what communicative challenges does the team have? What ends might be served by making implicit commitments explicit? How might insights be captured for later use by the team? Answers to these questions are a function of the specific concerns for the team. If it has encountered challenges in its process, TDI will determine if there are ways in which a dialogue might help. For example, if team members have talked past one another because of lack of familiarity with the technical vocabulary of the constituent disciplines, a Toolbox workshop might help reveal those terms and their use in the project. If the team is too new to have encountered any difficulties, a workshop can illuminate the conceptual and normative determinants of the thinking of its members, thereby positioning itself to avoid potential snags and leverage complementary strengths. Teams might stabilize the insights from their discussions by creating a team glossary. The creation of such a glossary is a possible Toolbox co-creation activity.

Partner goals in workshops we've run have included building research capacity, identifying relevant stakeholders, fostering new collaborations, exploring power dynamics within the team, defining mission, vision, and goals, and developing next steps for team action. The attainment of these workshop goals requires critical reflection and analysis, identification of team and individual values and priorities, and reflection on previous team process. Often, the overall goal of a workshop can be understood as building collaborative and communicative capacity.

For an example of this process, consider the steps that led to the RIHN workshop. In 2015, TDI was contacted by Dr. Steven McGreevy, an environmental sociologist at the Research Institute for Humanity and Nature (RIHN) in Kyoto, Japan. He asked TDI to run capacity-building workshops for several research teams. The initial phase of interaction between Dr. McGreevy, RIHN, and TDI involved email and virtual exchanges that focused on identifying what the workshops might achieve and what sort of budget would be appropriate. RIHN recognized the value of capacity-building workshops and had conducted them in the past for their research teams, and Dr. McGreevy thought that TDI could provide another helpful workshop experience along those lines. Together, Dr. McGreevy, RIHN, and TDI identified three objectives that a Toolbox workshop could achieve at RIHN. These objectives were to improve the research capacity of teams by providing resources to identify obstacles to effective communication and collaboration, to enhance communication and increasing collaborative capacity by reducing the amount of information "lost in translation" between different disciplines, and to identify aspects of each research project where there could be differences in values and beliefs.

2.3.2 Workshop Protocol Design

Once goals have been identified, TDI works with the partner to develop a workshop protocol to achieve those goals. Such a protocol specifies three elements: (1) the design of the workshop itself, (2) the Toolbox instrument that structures the workshop's dialogue, and (3) a co-creation activity, which is included to help participants convert insights achieved in dialogue into concrete project gains. Specification of workshop design includes determination of when and where the workshop will be held, whether it will be in-person or virtual, and what its duration should be. Specification of the Toolbox instrument design requires work prior to the workshop, in collaboration with the partner, to ensure that the dialogue addresses the issues that matter for the team. TDI has used a variety of mechanisms to gather information from people who will participate in the workshop that help clarify the issues that should be the focus of the dialogue. These mechanisms have included surveys, interviews, and mental models.

For example, workshop participants might complete a brief survey that solicits their perspectives on challenges faced by the team. Together with the team, TDI uses this information to identify the issues about which there may not be full understanding. A selection of these issues constitutes the *core questions* that introduce the module themes in the instrument. TDI then works with the team leaders or their designates to develop Toolbox prompts to articulate aspects of those issues, with those prompts constituting the Toolbox instrument used to structure the dialogue.

Returning to the specific example of the RIHN workshop, development of the RIHN Toolbox protocol took place over the course of three months. After several discussions and the presentation by TDI of a proposed protocol, RIHN requested a workshop structure that included an initial lecture, an ice-breaker exercise, the standard Toolbox dialogue sessions, and the co-creation activity of a "build-your-own" Toolbox session that included a concept-mapping exercise to be conducted by each team (cf. O'Rourke et al. 2018) (see Table 2.1).

Dr. McGreevy and TDI also developed a RIHN-focused Toolbox instrument for the morning dialogue sessions. This began with the completion of a preliminary worksheet that provided information about the challenges to collaboration, communication, and success that RIHN faces, what RIHN might gain from the Toolbox experience, who the participants would be, and the logistical or institutional constraints that were relevant to the workshop (e.g., whether anyone would be participating virtually and the sizes of the workshop rooms). There was also an exchange of research publications that provided Dr. McGreevy and TDI with clarity about each other's work. Additionally, TDI conducted a preliminary survey of researchers who planned to participate in the workshop about their familiarity with their collaborators

---

TABLE 2.1
Plan for RIHN Toolbox Workshop

Day One—RIHN Toolbox Workshop
- Opening TDI lecture to the entire community (60–90 minutes)

Day Two
- Preamble to Toolbox workshop—plenary (30 minutes)
- Small group standard Toolbox sessions—breakout by research group (120 minutes)
- Debrief discussion of the Toolbox sessions—plenary (30 minutes)
- *LUNCH*
- Report from the Toolbox Team about the workshops from the morning (30 minutes)
- Discussion of report (30 minutes)
- Build-your-own-Toolbox session—real-time philosophical evaluation of their problem space (120 minutes)
  - Concept-mapping exercise of their project
  - Identification of different values and priorities with respect to the project
  - *BREAK*
  - Development of prompts that articulate the differences among them
  - Discussion of the prompts
- General discussion/debrief and next steps (60 minutes)

and project stakeholders, their objectives and values as environmental researchers, and the character of communication in the project. This information, together with an exchange of ideas via email, led to the development of a Toolbox instrument that contained these thematic modules and their associated core questions: motivation for conducting research, methodology, confirmation, reality, values, and transdisciplinary research. Once the instrument was finished, the prompts were translated into Japanese for the workshop. See Table 2.2 for the prompts organized by module.

TABLE 2.2
RIHN Toolbox Instrument Questions and Statements (Each Statement a Rating-Response Item with a 1–5 Disagree-Agree Scale and "I Don't Know" and "Not Applicable" Also as Options)

RIHN Toolbox Instrument

Motivation

Core Question: Does the principal value of research stem from its applicability for solving problems?

1. The principal value of research stems from the potential application of the knowledge gained to real-world problems.
2. Solving stakeholder problems should be the primary objective of RIHN research projects.
3. My disciplinary research primarily addresses basic questions.
4. RIHN projects are as much about basic science as they are about applied science.
5. The members of the RIHN community have similar views concerning the motivation core question.

Methodology

Core Question: How should disciplinary methods figure in transdisciplinary research?

1. Scientific research must be hypothesis driven.
2. Establishing harmonious human-environmental relations requires reliance on the methods of social science.
3. In order for environmental research to be valid, it must be quantitative.
4. High-quality interdisciplinary research requires collaborators to understand all methods used.
5. Scientific research can be valid without being experimental.
6. To address global environmental challenges, sustainability science must incorporate methods from the humanities.
7. The members of the RIHN community team have similar views concerning the methodology core question.

Confirmation

Core Question: What types of evidentiary support are required for knowledge?

1. Full validation of a result in environmental science requires that it be applied by stakeholders.
2. There are strict requirements for determining when empirical data confirm a tested hypothesis.

TABLE 2.2

RIHN Toolbox Instrument Questions and Statements (Continued)

3. Validation of results requires replication.
4. Unreplicated results can be validated if confirmed by a combination of several different methods.
5. Research interpretations must address uncertainty.
6. The members of the RIHN community have similar views concerning the confirmation core question.

Reality

Core Question: Do the products of scientific research more closely reflect the nature of the world or the researchers' perspective?

1. Scientific research aims to identify facts about a world independent of the investigators.
2. Scientific claims need not represent objective reality to be useful.
3. Models invariably produce a distorted view of objective reality.
4. The subject of my research is a human construction.
5. Adequate treatment of environmental problems requires a global response.
6. The members of the RIHN community have similar views concerning the reality core question.

Values

Core Question: Do values negatively influence scientific research?

1. Objectivity implies an absence of values by the researcher.
2. Incorporating one's personal perspective in framing a research question is never valid.
3. Value-neutral scientific research is possible.
4. Determining what constitutes acceptable validation of research data is a value issue.
5. Allowing values to influence scientific research is advocacy.
6. Sponsored research is biased research.
7. The members of the RIHN community have similar views concerning the values core question.

Reductionism

Core Question: Can the world under investigation be reduced to independent elements for study?

1. Differences in spatiotemporal scales impede useful synthesis in cross-disciplinary research.
2. The world under investigation is fully explicable in terms of its constituent parts.
3. The world under investigation must be explained in terms of the emergent properties arising from the interactions of its individual components.
4. My research typically isolates the behavior of individual components of a system.
5. Scientific research must include explicit consideration of the context in which it is conducted.
6. The members of the RIHN community have similar views concerning the reductionism core question.

(continued)

TABLE 2.2
RIHN Toolbox Instrument Questions and Statements (Continued)

Transdisciplinary Research

Core Question: How should non-academic stakeholders be involved in socio-environmental research?

1. When scientific and traditional knowledge conflict, priority should always be given to traditional knowledge.
2. Transdisciplinary research cannot be successful without democratic participation of stakeholders.
3. Sometimes non-academic stakeholders are just wrong about what is in their best interests.
4. Academic researchers have as much to learn from stakeholders as stakeholders do from academic researchers.
5. Stakeholders should have input into the methods used in a transdisciplinary research project.
6. The members of the RIHN community have similar views concerning the transdisciplinary research core question.

2.3.3 THE WORKSHOP EXPERIENCE

Prior to a workshop, participants are typically asked to read a TDI article that describes the workshop and its purpose. We also ask potential participants to review TDI protocols for the workshop so that their participation in the workshop is fully informed. (Data collection during Toolbox workshops is covered by US and international rules concerning human subjects research.) At the start of a workshop, the workshop facilitator delivers a *preamble* that briefly introduces the Toolbox dialogue method, articulates the goals for the session, and describes how the workshop is designed to achieve those goals. If the participants are unfamiliar with one another, the facilitator introduction will typically be preceded by participant introductions.

The participants then divide into dialogue groups. These usually have no more than 12 members, and they are typically facilitated by a member of the TDI team. Each participant completes the Toolbox instrument, scoring each prompt using a five-point rating-response scale that is like a Likert scale. This scale ranges from "disagree to agree" with additional "I don't know" and "Not applicable" options. The prompts, together with their responses, serve as the focus of the subsequent facilitated dialogue, which can last as long as two hours but typically lasts one hour or so. Once the dialogue has concluded, the participants complete the instrument a second time. Depending on the amount of time allotted, the dialogue might be followed immediately by a report out and debrief, which is especially important for large groups that are divided into multiple parallel dialogue sessions.

Following a break, the workshop then moves into a co-creation activity. The co-creation activity is designed to leverage what is learned in the dialogue portion of the workshop with the aim of creating a stable, public product that participants can take away. If the participants are a team, the product is intended to be an input into their ongoing collaboration. This activity serves to ensure that the gains made through reflexivity and perspective seeking are not lost. The co-creation activity typically begins with a preamble to describe the exercise, after which the participants are divided into groups of three or four to work on specific activities, such as a concept map of their part of the project, a glossary of technical terms used in the project, or a specification of top priorities for the project. This part of the workshop ends with time for reporting out and a debrief discussion of the full experience.

Toolbox workshops typically last for no more than four hours, but the RIHN workshop began with an evening lecture on transdisciplinary research and was followed by a full-day workshop the next day. The second day began with a preamble and a pair of standard Toolbox dialogue sessions. One unprecedented aspect of this workshop was that some of the dialogue was in Japanese; this limited the facilitators' ability to understand the dialogue and work with the data after the workshop.

After lunch, the workshop commenced with a reflective report by TDI about the morning sessions. This included notes from the pre-workshop survey of participants, an accounting of the prompts discussed in the dialogues, and highlights from those parts of the dialogues the facilitators were able to track. The highlights included discussions of the relationships between qualitative and quantitative methods, consideration of the differences between the humanities and the social sciences, and affirmation that since collaborators work with *people* and not *disciplines*, it is important not to assume that your collaborators will conform with any stereotype you have for their discipline.

The focus of the afternoon session shifted to the specific research topics of the teams in attendance. Representatives of nine teams participated in the workshop. Participating teams included the "Lifeworlds of sustainable food consumption and production: Agrifood systems in transition" team and the "Mathematical-geographical modeling on divergencies of humanity and nature in early and pre-modern worlds" team. If present, multiple members of a team gathered to produce a concept map of their project, which conceptualized the project in terms of stocks and flows (cf. Heemskerk et al. 2003), noting where differences in values and beliefs among the researchers and stakeholders could arise. They were then asked to articulate these differences in the form of Toolbox-style prompts which could be used to reveal project junctures where problems could arise. The session ended with a report from each group about this co-creation activity.

2.3.4 WORKSHOP FOLLOW-UP

Post-workshop activities involve additional data collection, data processing by the TDI team (see below), and data sharing between TDI and the participant team. This last activity involves both discussions of what was learned during the workshop and what future activities the team and TDI might engage in, which could include further collaboration with TDI itself. We now discuss these processes and highlight elements that are conducted with the participants.

Within two weeks after a workshop, participants are invited to provide feedback on the Toolbox workshop via an online questionnaire. Participants are asked if they

TABLE 2.3

STEM Workshop Participant Assessments of the Impact of the Toolbox Workshop (Schnapp et al. 2012, p. 471)

Key Themes from Open-Ended Responses to Post-Workshop Evaluations	Percent
Workshop had a positive impact on awareness of the knowledge, opinions, or scientific approach of teammates	84.9
Overall assessment was entirely positive	82.6
Statements about impact on professional development were entirely positive	77.4
Workshop helped participant become more aware of dimensions of cross-disciplinary research, including challenges associated with working across disciplines and awareness of other disciplinary perspectives	43.9
Workshop helped participant become more aware of dimensions of science or scientific research	41.7
Workshop had (or could have) a positive impact on research communication	33.1
Workshop had a positive impact on the social aspects of team building	18.7
Made at least one skeptical or negative comment about some aspect of the Toolbox workshop	8.6

enjoyed the workshop, what the characteristics of the conversation were (e.g., "Our conversation was an open exchange of thoughts and ideas"), what impact it had on their understanding (e.g., "I have a better understanding of my colleagues' attitudes toward communication and actionable research because of the Toolbox dialogue"), and what impact it had on their group (e.g., "Participating in the Toolbox workshop has had a positive influence on our team"). These statements are associated with a rating-response scale that structures participant responses. The questionnaire ends with open-ended questions about key insights, any under-discussed prompts, and missing topics. Table 2.3 presents an analysis of responses to such a questionnaire (Schnapp et al. 2012).

Questionnaire data and data collected prior to and during the workshop inform the report that TDI writes for the partner team. These data typically include the following: pre-workshop conversations and questionnaire responses from team leaders about objectives and core themes; pre-workshop survey of workshop participants, as described above; pre-dialogue and post-dialogue rating-response data from completed Toolbox instruments; audio recording and transcription of the workshop dialogue; products generated during the co-creation activity; and post-workshop questionnaire data from the participants.

The reporting phase of the RIHN workshop was unusual in several ways. First, the TDI representatives produced a preliminary analysis of the results of the morning session and communicated their results immediately after lunch. Second, Dr. McGreevy had reporting requirements within RIHN and took responsibility for producing a report from the experience, so he did not rely upon TDI to produce a report. TDI evaluated the data collected during the RIHN workshop; in addition, a post-workshop questionnaire was translated and made available to the RIHN participants online. TDI prepared summaries for RIHN of the data from the session

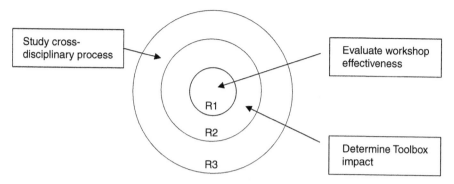

FIGURE 2.2 The three nested levels of TDI research as informed by data collected in Toolbox workshops and subsequent reflection on the Toolbox workshop experience.

and delivered de-identified responses to the Toolbox post-workshop questionnaire. Finally, TDI recommended that RIHN use Toolbox prompts in future reflexive dialogues.

2.3.5 FROM REPORT TO RESEARCH

We consider the report to be part of the research TDI conducts in the wake of each Toolbox workshop. TDI conducts research related to Toolbox workshops on three nested levels—the workshop level (R1), the Toolbox approach level (R2), and the level of cross-disciplinary process (R3). (See Figure 2.2.) The levels are mutually informative. What we learn at the level of the workshop informs our understanding of how the Toolbox approach functions in general, e.g., close attention to the specific character of individual workshops gives us insight into which conditions are good predictors of Toolbox success. Since the Toolbox approach is used in a variety of contexts to enhance cross-disciplinary capacity, a better understanding of how it functions across these contexts can deepen our understanding of cross-disciplinary dynamics more generally. Finally, research into the more general aspects of cross-disciplinary process and practice feeds back into our thinking about specific workshops, providing us with information that we can use to ensure that the next partner team is well-served by the experience.

Research at level R1 centers on the analysis of workshop data in order to identify strengths and weaknesses in the communication and collaboration dynamics of the participating team. This evaluation can identify troublesome areas where additional conversation could be helpful as well as fruitful areas for future collaboration. The results of this evaluation are included in a report that specifies workshop results from TDI's perspective. In addition to a summary of the workshop as a whole and the path taken in dialogue by the participants, the report provides descriptive statistics of the rating-response data and recommendations for the participating team. The recommendations highlight the strengths and weaknesses revealed by data analysis and leverage the experience of TDI in running over 300 of these workshops. These reports are shared with the team and TDI helps with interpretation.

Research at level R2 involves comparison of data across the workshops to assess the impact of the Toolbox approach. This is valuable internally as a guide to improving the Toolbox approach and externally to those interested in the Toolbox approach as a capacity-enhancing intervention. Research at this level has been published in journals across a range of disciplines (Schnapp et al. 2012; O'Rourke and Crowley 2013; Gonnerman et al. 2015).

Research at level R3 involves reflection on Toolbox experiences with a view to evaluating research questions that concern cross-disciplinary process in general. Work at this level can be empirical, but it is often focused on philosophical questions that concern cross-disciplinary research. TDI has used data from this level to investigate communication and collaboration (O'Rourke et al. 2014), integration (O'Rourke et al. 2016), disagreement (Crowley et al. 2016), values in science (Steel et al. 2017), responsible conduct of research (Berling et al. 2019), and interdisciplinary reasoning (Laursen 2018). These investigations contribute to the feedback loop mentioned above by yielding insights that have helped us modify our understanding of cross-disciplinarity and what we think the Toolbox approach should enhance within cross-disciplinary projects.

2.4   A DEEPER DIVE INTO THE TOOLBOX WORKSHOP EXPERIENCE

Buy-in from the participants is important to the success of a Toolbox workshop, and so it is crucial that a team participating in a workshop regards the experience as valuable. There can be barriers to such buy-in, however. Time is precious, and the value of activities that don't contribute *directly* to a project can be easy to miss. This problem can be exacerbated if the nature and purpose of those activities are misunderstood. Below, we correct some common misunderstandings regarding the workshop and comment on the value of the workshop to team success. We begin with a brief account of how the workshop has evolved and then discuss three central elements of the workshop experience

2.4.1   TOOLBOX 2.0

We think of the current workshop process, from initial team engagement through co-creation activity and reporting, as "Toolbox 2.0." We contrast this with "Toolbox 1.0." (Chapter 3 provides details on the historical transition from the analogue phase of Toolbox 1.0 to the digital phase of Toolbox 2.0.) Toolbox 1.0, which was presented first by Eigenbrode et al. (2007) and later elaborated by Looney et al. (2014), was an explicitly philosophical, top-down approach that emphasized the epistemological and metaphysical aspects of scientific research. The idea was that there are philosophical dimensions of scientific research common to a wide range of disciplines that can be used to map out the space of similarities and differences among collaborators. Development of this approach yielded a Toolbox instrument—the Scientific Research Toolbox instrument (see Appendix A)—that was used to structure dialogue in most of the early workshops (Looney et al. 2014). What this meant in practice was that a team would participate in a dialogue and then engage in a debrief discussion.

Information about the team was gathered in advance to inform the facilitator. This information influenced the preamble and how the dialogue groups were determined; however, it did not change the instrument. Co-creation activities were not included as part of Toolbox 1.0. There was no intent to forge longer-term working relationships with TDI.

Since 2011, we have focused on designing workshops from the bottom up. A Toolbox 2.0 workshop begins by having a team imagine with TDI what a successful Toolbox workshop experience would look like for them, and then they work together to reverse-engineer a Toolbox protocol to make that experience likely. The resulting workshops include instruments that are constructed to reflect the specific interests of the team and can be organized in ways that are more suitable to a particular environment (e.g., conducted virtually rather than face-to-face). They also typically include a co-creation activity designed to leverage what is learned from the dialogue in creating a tangible product that can be used in the future work of the team. In other words, we meet teams where they are at in Toolbox 2.0, making it more likely that they become interested in entering into a longer-term working relationship with TDI. This approach arises from more than a decade of experience, through which we have learned that one size doesn't fit all when it comes to structured dialogue and that such a dialogue has more value if it is responsive to a team's specific needs.

2.4.2 THE TOOLBOX INSTRUMENT

The Toolbox instrument is a key part of any Toolbox workshop. Prior to a workshop, TDI works with team representatives to identify three to six central themes and to formulate statements that express aspects of those themes. The themes become the focus of "modules," which comprise a "core question" that articulates the theme and five to eight prompts that express aspects of the theme. Each prompt is associated with a 1–5, Disagree-Agree rating-response scale and options for "I don't know" and "Not applicable." A Toolbox instrument is a set of modules. Taken together, they focus workshop participants on issues and commitments that can be used in dialogue to map out the similarities and differences among team members.

For example, consider the Transdisciplinary Research module in the RIHN Toolbox Instrument (Figure 2.3). This module concerns the participation of non-academic stakeholders in research projects, with a specific focus on socio-environmental research. Information gathered before the workshop indicated that collaborating with non-academic stakeholders was a key challenge for the RIHN teams that would be working with us. Inclusion of this module allowed us to structure dialogue about important considerations such as traditional knowledge, democratic participation, and the role of stakeholders in complex research projects.

The prompts in an instrument are expressions of a way of thinking about possibly contentious core beliefs and values, and the rating-response scale is used to encourage a reaction by participants to that way of thinking, relative to their specific (and quite possibly different) interpretations. For example, in the Transdisciplinary Research module in Figure 2.3, the fourth prompt asserts that academic researchers are not in a superior epistemic position to stakeholders but, rather, have as much to learn as they have to teach. This provocative statement is used to identify (both to

RIHN Toolbox Instrument

Transdisciplinary Research

Core Question: How should non-academic stakeholders be involved in socio-environmental research?

1. When scientific and traditional knowledge conflict, priority should always be given to traditional knowledge.

    Disagree   Agree
       1   2   3   4   5   I don't know   N/A

2. Transdisciplinary research cannot be successful without the democratic participation of stakeholders.

    Disagree   Agree
       1   2   3   4   5   I don't know   N/A

3. Sometimes non-academic stakeholders are just wrong about what is in their best interests.

    Disagree   Agree
       1   2   3   4   5   I don't know   N/A

4. Academic researchers have as much to learn from stakeholders as stakeholders do from academic researchers.

    Disagree   Agree
       1   2   3   4   5   I don't know   N/A

5. Stakeholders should have input into the methods used in a transdisciplinary research project.

    Disagree   Agree
       1   2   3   4   5   I don't know   N/A

6. The members of the RIHN community have similar views concerning the Transdisciplinary Research core question.

    Disagree   Agree
       1   2   3   4   5   I don't know   N/A

FIGURE 2.3 The "Transdisciplinary Research" module from the RIHN Toolbox instrument. A Japanese translation of this instrument was also produced for this workshop.

themselves and to the group) those who prize academic expertise and those who are more skeptical of it, and this identification then serves to highlight a difference worth investigating in dialogue.

The prompts are written to reflect a variety of different perspectives, e.g., prompts invite participants to think about their project, their discipline, their team, and even themselves. (For more on this, see Robinson et al. 2019.) In the Transdisciplinary Research module in Figure 2.3, the first five prompts articulate different claims about the transdisciplinary context that emphasize non-academic stakeholders. Most of these advocate an expanded stakeholder role, whereas one of them (the third prompt) invokes caution about an expanded role. These prompts are written from an abstract, third-person perspective that can be easily adopted by a transdisciplinary scientific researcher, regardless of the context of their work. The last prompt, by contrast, asks the participant to reflect on the collective attitude within the RIHN community.

An important challenge for participants in cross-disciplinary projects is the use of terms in different ways (Wear 1999; Bracken and Oughton 2006; Thompson 2009). One benefit of a Toolbox workshop is the realization that terms held in common, such as "hypothesis," "model," and "applied," have different interpretations across disciplines (Donovan et al. 2015). Since these terms have different interpretations, participants in Toolbox workshops often ask us what we mean by them, e.g., "what do we mean by 'values'?" Or "what do we mean by 'evidence'?" The technical terms in the prompts are typically used by the participants themselves, but facility in using a term does not necessarily translate into an ability to define it, especially when the terms are encountered outside of typical research contexts.

A request for a definition in these situations makes perfect sense, given that the terms involved admit of different interpretations. Workshop facilitators, however, reject such requests, directing the participants to compare definitions that strike them as reasonable given their disciplinary perspectives. In issuing this direction, facilitators are motivated by the idea that workshop participants can only come to realize that terms admit of different interpretations across disciplines if they use them as they normally do in their respective home disciplines. Through the process of discussing, debating, and occasionally disagreeing on possible definitions, participants often learn something about how they use the terms in their research practice, in addition to learning about the range of meanings that are possible for these important terms and disciplinary differences in how the participants think about the issues articulated by the prompts. It is for these reasons that we don't supply a glossary of terms in the workshop.

From a more technical perspective, participant desire for a glossary is often motivated by the fact that prompts contain terms that are *vague* or *ambiguous*. A prompt is *vague* when it contains a term of central importance that admits of borderline cases between those where the term clearly applies and those where it clearly does not. For example, the second prompt in Figure 2.3 employs the term "successful" in connection with transdisciplinary research, which is not clearly applicable across a wide range of cases. In contrast, an *ambiguous* prompt contains a term that has multiple relevant meanings. Examples are "input" and "methods" in the fifth prompt. Toolbox workshop prompts often deliberately use vague or ambiguous language. If the point of these prompts was to collect survey data, they would thereby be poorly designed, since one cannot be sure that vague or ambiguous prompts are interpreted in the same way and so one arguably cannot meaningfully compare responses. Toolbox workshop prompts, however, are not intended to measure attitudes—they are not psychometrically validated survey prompts.

The point of workshop prompts, instead, is to provoke responses and generate interesting conversation. They are meant to get people talking. They are also invitations to participants to modify the prompts in dialogue so that they reflect views that are more in line with their thinking. Typically, a vague prompt generates uncertainty among the participants about how it should be interpreted. The ensuing dialogue can reveal important similarities and differences among the participants if uncertainty fuels an examination of the difference between clear-cut cases and borderline cases related to the prompts. For example, the second prompt in Figure 2.3 might structure dialogue about what counts as *successful* transdisciplinary research, which could clarify how participants view their common project.

Ambiguous prompts can also generate productive uncertainty. For example, the fifth prompt in Figure 2.3 contains the term "input," which can provoke participants to wonder what types of input are relevant and thereby reveal differences of opinion among collaborators about what counts as input. It could easily turn out that collaborators have a very different understanding of what kinds of input they should contribute. This highlights the value of dialogue as a way of locating and coordinating differences in technical vocabulary.

There is one other rhetorical aspect of Toolbox prompts designed to stimulate dialogue, namely, the use of extreme terms such as "always" and "must," which can appear to make a prompt too easy to accept or reject. For example, the first prompt in the Transdisciplinary Research module might seem easy to disagree with, because it is easy to imagine situations in which scientific knowledge is as important as traditional knowledge. However, the prompt highlights an important issue: under what circumstances is it appropriate to give priority to scientific knowledge when it conflicts with traditional knowledge? The extreme character of this prompt underwrites a dialogue that can enable the group to move past strongly divided opinions about the priority of different types of knowledge. This reflects the fact that—once again—the Toolbox instrument is *not* a survey instrument, since prompts containing extreme terms won't function as useful tools for teasing apart different attitudes in a sample. Instead, it is an instrument for structuring dialogue about the beliefs and values that frame one's research practice.

2.4.3 TOOLBOX DIALOGUE

The dialogue is the defining feature of the Toolbox workshop. It is a vehicle for collective self-discovery, enabling workshop participants to learn things about themselves that enhance their mutual understanding. It is the principal vehicle for capacity-building that TDI employs. Philosophy structures the dialogue in that the Toolbox prompts draw on philosophical concepts and methods to reveal commitments that frame how individuals understand research problems. (For more on this, see Chapter 6.) These commitments reveal aspects of research problems that an individual finds problematic and delineate the space of options for dealing with the problems. The workshop allows each participant to react to prompts in ways that reveal these commitments. The listening and reciprocity that mark substantive dialogue put the team in a position to work with these commitments so as to aid progress toward project goals.

The Toolbox dialogue can be viewed as a teaching and learning mechanism for the team. Each participant is a *teacher* of their own view of the prompts and the team project. Each participant is also a *learner* since each acquires information about how their teammates think and act. Thus, the dialogue supports both individual and team reflexivity (Goodyear and Zenios 2007; Tsoukas 2009), which in addition to enhancing mutual understanding can contribute to a culture of reflection that helps the team identify when the project is going well and when it is going poorly (for more on this, see Chapter 11).

In a Toolbox workshop, the dialogue creates an environment in which the team considers its own beliefs, values, interests, and priorities. We encourage the team to make the dialogue their own by beginning with any prompt they choose and following their own trajectory through the instrument. Letting the participants determine the initial topic positions the team to determine the initial conditions of this trajectory, often selecting a prompt that has special meaning for them or stands out for one of their members. Once the dialogue begins, we want the participants to follow their interests around the instrument—there is no set order to discussion of the prompts. While many workshop groups talk about the prompts, using the rating responses to compare attitudes, this way of engaging with the instrument isn't required. Occasionally participants focus on the core questions, or even talk at a higher level of abstraction about the general themes of the modules. By allowing the group to determine their own way through the instrument, we learn about how they engage with the issues it raises and how they communicate with one another.

If TDI facilitators imposed a path through the instrument, it would make the dialogue more about their interests. As facilitators, we work to ensure that a team visits all the modules, but we don't require that a team discuss all prompts. In fact, which prompts are discussed and which are not helps us identify issues that stand out to the team as most interesting or most relevant, and also areas of difference and similarity among participants that are underappreciated by the team. On occasion, the facilitators allocate time to each module and designate time to wrap up loose ends, but usually they let the participants determine how they want to use their time and where to focus their attention.

Facilitators do not demand that all participants speak during dialogue sessions, although we hope this occurs. There is evidence that teams with an even balance of speaking turns in meetings are more effective and successful (Duhigg 2016). Facilitators emphasize this finding in the preamble to the workshop. Of course, not everyone is equally willing to contribute to the dialogue, and this hesitance is exacerbated for some by the fact that Toolbox dialogue is more abstract and further afield from research conversations they find typical. To some extent this unfamiliarity is offset by the fact that our prompts are written so that they do not have right answers, which is another point we emphasize in our preamble. The prompts are written to express points of view that are grounded in contingent ways of thinking about fundamental features of research or practice. For example, the Scientific Research Toolbox instrument (see Appendix A) includes the prompt, "Scientific research (applied or basic) must be hypothesis driven," which turns out not to be a universal way of thinking about how the scientific process works (Donovan et al. 2015).

2.4.4 THE CO-CREATION ACTIVITY

The final element of the Toolbox 2.0 workshop is the co-creation activity. Its goal is to create a stable, public, and team-relevant product that can be used by the team in pursuing project objectives. Insights achieved during dialogue can be ephemeral. If, however, the dialogue generates a tangible product that is relevant to the team, then the value of dialogue as a contribution to team process is reinforced.

Co-creation activities vary among groups. Examples can be found in Lipmanowicz and McCandless (2014) or on their website (www.liberatingstructures.com/). The website focuses on what they call "liberating structures," which are approaches to collaboration and associated activities that "enhance relational coordination and trust." Such activities can be more structured (e.g., concept mapping or the production of a project glossary) or less structured (e.g., collective brainstorming or think-pair-share). Co-creation activities in the Toolbox context often begin with individual reflection and writing, followed by small group work on the selected activity. This group might be as small as two in a think-pair-share activity or as large as six, but typically we have three to four people in the co-creation breakout groups. We plan for the co-creation activity in advance, working with team leadership to identify an exercise that would strike the team as useful; however, we remain flexible so that on-the-spot changes can be made that will make the activity more valuable.

2.5 CONCLUSION

Collaborating with TDI entails the creation of a bespoke workshop designed specifically for the partner team according to a bottom-up process that involves structured dialogue and a coupled co-creation activity. These workshop elements should reveal implicit beliefs and values that influence how collaborators deliberate and make decisions about their common research project. This in turn enables teams to boost performance by leveraging differences and similarities among these implicit commitments to advance the team toward its project objectives.

2.6 ACKNOWLEDGMENTS

The Toolbox dialogue method is a dynamic method that evolved over time, thanks to the contributions of all those present and former Toolboxers who have contributed to the effort. We are very grateful to all of them but especially to Brian Robinson who was a key part of making the RIHN Toolbox workshop a success. We would also like to think Steven McGreevy of RIHN for his interest in the Toolbox dialogue method and for being such a great collaborator and host. Thanks are also due to Graham Hubbs and Steven Orzack for their editing assistance. O'Rourke's work on this chapter was supported by the USDA National Institute of Food and Agriculture, Hatch Project 1016959.

2.7 LITERATURE CITED

Berling, E., C. McLeskey, M. O'Rourke, and R. T. Pennock. 2019: A new method for a virtue-based responsible conduct of research curriculum: Pilot test results. Science and Engineering Ethics 25:899–910.

Bracken, L. J., and E. A. Oughton. 2006: "What do you mean?" The importance of language in developing interdisciplinary research. Transactions of the Institute of British Geographers 31:371–382.

Crowley, S. J., C. Gonnerman, and M. O'Rourke. 2016: Cross-disciplinary research as a platform for philosophical research. Journal of the American Philosophical Association 2:344–363.

Donovan, S. M., M. O'Rourke, and C. Looney. 2015: Your hypothesis or mine? Terminological and conceptual variation across disciplines. SAGE Open 5:1–13.

Duhigg, C. 2016: What Google learned from its quest to build the perfect team. The New York Times Magazine.

Eigenbrode, S. D., M. O'Rourke, J. D. Wulfhorst, D. M. Althoff, C. S. Goldberg, K. Merrill, W. Morse, M. Nielsen-Pincus, J. Stephens, L. Winowiecki, and N. A. Bosque-Pérez. 2007: Employing philosophical dialogue in collaborative science. BioScience 57:55–64.

Fiore, S. M. 2008: Interdisciplinarity as teamwork. Small Group Research 39:251–277.

Gonnerman, C., M. O'Rourke, S. J. Crowley, and T. E. Hall. 2015: Discovering philosophical assumptions that guide action research: The reflexive toolbox approach. Pp. 673–680 in H. B. Huang and P. Reason, eds. The SAGE Handbook of Action Research, 3rd edition. SAGE, Thousand Oaks, CA.

Goodyear, P., and M. Zenios. 2007: Discussion, collaborative knowledge work and epistemic fluency. British Journal of Educational Studies 55:351–368.

Heemskerk, M., K. Wilson, and M. Pavao-Zuckerman. 2003: Conceptual models as tools for communication across disciplines. Conservation Ecology 7:8.

Hoever, I. J., D. Van Knippenberg, W. P. Van Ginkel, and H. G. Barkema. 2012: Fostering team creativity: Perspective taking as key to unlocking diversity's potential. Journal of Applied Psychology 97:982–996.

Laursen, B. K. 2018: What is collaborative, interdisciplinary reasoning? The heart of interdisciplinary team science. Informing Science 21:75–106.

Lewis, K., M. Belliveau, B. Herndon, and J. Keller. 2007: Group cognition, membership change, and performance: investigating the benefits and detriments of collective knowledge. Organizational Behavior and Human Decision Processes 103:159–178.

Lipmanowicz, H., and K. McCandless. 2014: The Surprising Power of Liberating Structures: Simple Rules to Unleash a Culture of Innovation. Liberating Structures Press, Seattle.

Looney, C., S. Donovan, M. O'Rourke, S. J. Crowley, S. D. Eigenbrode, L. Rotschy, N. A. Bosque-Pérez, and J. D. Wulfhorst. 2014: Seeing through the eyes of collaborators: Using Toolbox workshops to enhance cross-disciplinary communication. Pp. 220–243 in M. O'Rourke, S. J. Crowley, S. D. Eigenbrode, and J. D. Wulfhorst, eds. Enhancing Communication and Collaboration in Interdisciplinary Research. SAGE, Thousand Oaks, CA.

O'Rourke, M., and S. J. Crowley. 2013: Philosophical intervention and cross-disciplinary science: The story of the Toolbox Project. Synthese 190:1937–1954.

O'Rourke, M., S. J. Crowley, and C. Gonnerman. 2016: On the nature of cross-disciplinary integration: A philosophical framework. Studies in History and Philosophy of Science Part C: Studies in History and Philosophy of Biological and Biomedical Sciences 56:62–70.

O'Rourke, M., S. J. Crowley, S. Eigenbrode, and J. D. Wulfhorst. 2014: Introduction. Pp.1–10 in M. O'Rourke, S. J. Crowley, S. D. Eigenbrode, and J. D. Wulfhorst, eds. Enhancing Communication and Collaboration in Interdisciplinary Research. SAGE, Thousand Oaks, CA.

O'Rourke, M., T. E. Hall, J. Boll, B. Cosens, T. Dietz, J. Engebretsen, L. Goralnik, Z. Piso, S. Valles, and K. Whyte. 2018: *Values and responsibility in interdisciplinary environmental science: A dialogue-based curriculum for ethics education – Instructor kit.* Downloaded from http://eese.msu.edu/.

Okhuysen, G. A., and K. M. Eisenhardt. 2002: Integrating knowledge in groups: How formal interventions enable flexibility. Organization Science 13:370–386.

Robinson, B., C. Gonnerman, and M. O'Rourke. 2019: Experimental philosophy of science and philosophical differences across the sciences. Philosophy of Science 86:551–576.

Salazar, M. R., K. Widmer, K. Doiron, and T. K. Lant. 2019: Leader integrative capabilities: A catalyst for effective interdisciplinary teams. Pp. 313–328 *in* K. L. Hall, A. L. Vogel, and R. T. Croyle, eds. Strategies for Team Science Success: Handbook of Evidence-Based Principles for Cross-Disciplinary Science and Practical Lessons Learned from Health Researchers. Springer, New York.

Schnapp, L. M., L. Rotschy, T. E. Hall, S. J. Crowley, and M. O'Rourke. 2012: How to talk to strangers: Facilitating knowledge sharing within translational health teams with the Toolbox dialogue method. Translational Behavioral Medicine 2:469–479.

Steel, D., C. Gonnerman, and M. O'Rourke. 2017: Scientists' attitudes on science and values: Case studies and survey methods in philosophy of science. Studies in History and Philosophy of Science Part A 63:22–30.

Thompson, J. L. 2009: Building collective communication competence in interdisciplinary research teams. Journal of Applied Communication Research 37:278–297.

Tsoukas, H. 2009: A dialogical approach to the creation of new knowledge in organizations. Organization Science 20:941–957.

Wear, D. N. 1999: Challenges to interdisciplinary discourse. Ecosystems 2:299–301.

3 A Narrative History of the Toolbox Dialogue Initiative

Graham Hubbs

3.1 INTRODUCTION

The history of the Toolbox Dialogue Initiative (TDI) traces both a literal and a metaphorical movement from analogue to digital. In a literal sense, TDI has morphed from piles of paper to an array of online tools and digitized data, much like the transition from phonographic records to streaming music. In a metaphorical sense, although TDI looks quite different from what it did ten years ago, there is a unity underlying the differences that is centered on lightly facilitated, dialogue-based workshops organized around responses to Toolbox instruments. An LP is very different from an mp3, but they both do the same thing, viz., store acoustic information in a form that can be accessed electronically. Similarly, all versions of TDI, from the earliest to the most recent, do the same thing—they clarify concepts and uncover hidden assumptions in order to improve communication and collaboration.

The history of TDI begins long before anyone thought of structuring a philosophical conversation around a collection of statements evaluated by a five-point rating scale. It starts with an NSF-funded Integrative Graduate Education and Research Traineeship (IGERT) project at the University of Idaho (UI) that provided PhD students from a variety of agriculture and natural resource disciplines with cross-disciplinary, team-based training. In 2004, a professor, Sanford Eigenbrode, began to plan a seminar to investigate some of the philosophical underpinnings of science. The initial idea was to discuss and understand the philosophical dimensions of scientific validation addressed by Taper and Lele (2004). It soon became evident that the focus of that book was a part of a much broader set of issues concerning communication in science.

3.2 TOOLBOX BEFORE THE TOOLBOX: 2004–2007

Eigenbrode had identified *scientific inference* as a concept that could be used to compare and contrast disciplinary perspectives, and the Taper and Lele text seemed to offer a promising vehicle for exploring the concept. Eigenbrode began preparing the course with his student Chris Looney, and it was not long before they needed help. Looney had taken a few philosophy courses as an undergraduate and knew that forms of scientific inference are sometimes discussed by philosophers of science. He suggested enlisting a philosopher who could help and who might be interested in team-teaching the seminar. With the help of his fellow IGERT colleague, J. D. Wulfhorst, Eigenbrode was able to locate one: Michael O'Rourke. Early discussions between these three led to the conclusion that a narrow focus on the varieties of inference probably would not address the students' most serious needs. They needed a broad skill set that would enable them to collaborate in a cross-disciplinary environment. (We use "cross-disciplinary" as a generic term in place of more specific terms such as "multidisciplinary," "interdisciplinary," "transdisciplinary," and so on. See Eigenbrode et al. 2007 and Chapter 1.)

When the seminar began in 2005, students came from a variety of disciplines, including evolutionary biology, physics, a variety of social sciences, and philosophy. O'Rourke and Eigenbrode led the sessions, and Wulfhorst and Nilsa Bosque-Pérez, the director of the IGERT project, also attended. Despite being cross-listed between Philosophy and Entomology, no insects were ever discussed, and in spite of the original plan, not a single page of Taper and Lele's book was ever read by the class. After reading accounts of the challenges of cross-disciplinary collaboration, the group decided to focus on a topic that had been observed but not developed in detail, viz., the ways in which failures to communicate across disciplines can cause cross-disciplinary collaborations to collapse.

For an example of the sort of communication failure that interested them, imagine a new collaboration involving two biologists, one of whom is a laboratory-based biologist steeped in a frequentist tradition of statistical inference and the other a field ecologist who relies on Bayesian approaches to inference; further, imagine that neither is aware of this important difference, so that conversations about inference within our collaboration are marked by unrecognized misunderstandings. These collaborators may get by for a while, but it would not be surprising if one of them eventually makes a statement about inference that is unacceptable to the other. This could lead to the breakdown of the collaboration if they can't understand their differences and determine whether they can be reconciled. Similar difficulties can arise from differences in the perspectives of scientists that are grounded in other fundamental components of science such as reductionism and holism, the role of values in doing science, and even the very motivation for doing science. It became the research topic of the participants in this seminar to understand the nature of these difficulties.

The story so far demonstrates a basic lesson that guided TDI before it existed: be flexible. Eigenbrode had intended to probe a debate that was not likely to meet the needs of the students. The team restructured the seminar to make it more useful to the students, and then they let the students determine the topic that would be most helpful. Had the seminar stuck doggedly to Eigenbrode's original plan, there likely would be no TDI, and so none of the workshops, research, and collaborations that have flourished for more than a decade after the seminar.

The seminar participants read widely across the existing literature on communication in cross-disciplinary research, including Klein (1996, 2004), Kinzig (2001), Heemskerk et al. (2003), Stokols et al. (2003), Jakobsen et al. (2004), and Campbell (2005). They began to notice that despite the oft-repeated admonition to attend to communication in cross-disciplinary practice, there was little understanding of the level of communication required and how to achieve it. In order to develop such understanding, the team decided to run the rest of the seminar using the Pimentel method. In seminars run on this method, the first step is for the group to pick a research topic of the right scope to yield a publishable paper. The next step is to divide and conquer, each student taking on some portion of the project to become the expert in. Eigenbrode was exposed to this method when, as a student at Cornell University, he took a course taught by David Pimentel, who developed this way of teaching graduate students how to produce publishable and impactful research. Having identified a research topic, the seminar participants spent the semester acquiring expertise in aspects of the problem of cross-disciplinary communication and then were invited to continue their work together after the semester so that they could complete a paper describing their findings. This was published as Eigenbrode et al. (2007). The paper offered a solution to the problems of cross-disciplinary communication: the "toolbox for philosophical dialogue."

This toolbox was a set of questions designed to stimulate philosophical reflection on the nature of science. For TDI, this toolbox would prove to be analogous to Edison's first phonograph, tinfoil wrapped around a cylinder, still a step away from the flat disc form that would become ubiquitous as the long-playing analogue record. The intent behind the original toolbox was to stimulate dialogue among team members, who would share their answers and then engage in discussions of their views of scientific research. The questions in this toolbox included, "Should your collaborative research project emphasize applied over basic research?" "Can unreplicated results that are confirmed by a combination of methods qualify as knowledge?" "Can one integrate values into research and still remain objective?" and "Can subjective research be scientific?" It quickly emerged that these questions often prompted an answer to which almost all participants could agree—"It depends"—and this agreement proved difficult to explore in the lightly facilitated workshops. So long as workshop participants were unwilling to elaborate, this agreement meant that the workshops produced little revelatory dialogue about fundamental conceptual differences that might be impeding the team's work.

Like the improvements that produced the flat disc record after the original foil-covered cylinder, the toolbox was improved by replacing the probing questions with statements, each associated with a five-point rating scale akin to a Likert scale that could record participant agreement with the statements. (See Chapter 2.) The scale encouraged participants to commit to a more nuanced position on each statement. Participants were asked to share and defend their level of agreement with the prompts as part of the dialogue. This innovation proved effective in stimulating more revealing dialogue and in providing a new way of recording the spectrum of views held by participants in a workshop. The lesson from this innovation is to constantly look for what is not working and what can be done better. This complements the lesson about flexibility described above. If one finds, for example, that asking ambiguous and potentially vague philosophical questions is not sparking meaningful discussion, the solution to this problem might be to give up on asking questions altogether. With its new rating-scaled assertions, the philosophical toolbox had transformed into the Toolbox.

3.3   THE ANALOGUE YEARS I: 2007–2012

The step from an emerging method to a research project, from the toolbox method to the Toolbox Project (as it was known then), required what nearly all institutional science requires: money. In 2007, soon after the publication of Eigenbrode et al. (2007), O'Rourke and Eigenbrode pursued internal money at UI, which they obtained and used to run workshops with an eye to applying for external funding. The only data collected at this point were participant response data—participants were contacted after workshops and asked for their impressions of the experience. The response rate was not very high. Although the Toolbox team had included social scientists when it was working on Eigenbrode et al. (2007), they fell away in the period after that article's publication; in that period, when the Toolbox team was conducting exploratory workshops, they would have benefited from more engagement with social scientists in order to help understand the group dynamics they were engaged with and measure the effects of the workshop on these dynamics. The lesson here is that when doing cross-disciplinary science that has an important social dimension, talk to social scientists about your work, and if it seems they can help, get them actively involved.

The next move for O'Rourke, Eigenbrode, Wulfhorst, and Stephen Crowley, a philosopher at Boise State University, was to proceed with the data they had and to apply for funding through the National Science Foundation (NSF). Their proposal would take the Toolbox beyond UI to run workshops with IGERT teams across the nation. The proposal was well received, but funding was limited. Fortunately, they learned from UI's Office of Research and Economic Development that the state of Idaho was included within NSF's Experimental Program to Stimulate Competitive Research (EPSCoR), which supports scientific research in states with low federal research funding. EPSCoR funds were added in order to fund the Toolbox proposal, and the Toolbox Project had its first major external grant. The lesson here is that you don't know what you don't know. To prevent unknown unknowns from blocking progress, keep lines of communication open with those who are eager to offer support, financial or otherwise.

In 2008, Eigenbrode and O'Rourke took the Toolbox Project outside of Idaho for the first time and ran workshops at Cornell University. They received enthusiastic feedback. More importantly, they were asked two crucial questions. One: what were they doing with all of the Likert data they were collecting? Recall that the five-point rating scale was introduced to motivate more nuanced responses than "it depends" and not for research purposes; this question, though, motivated the Toolbox Project to expand the role played by participant answers to the rating scale. Second: were there plans to run several workshops with teams over a long span of time? This question prompted the Toolbox Project to look for long-term partners (for more on this, see Chapter 11). Over the course of the next year two further innovations became standard practice. Audio recordings were made of all workshops, which made conversational analysis of the transcripts possible, thus opening an additional line of Toolbox research. The standard workshop process was also modified so that the instrument would be administered at both the beginning and end of workshops. This allowed for assessing changes of opinion over the course of a workshop. During this time, the Toolbox team continued to hone the instrument. Under the guidance of

Shannon Donovan, probing assertions were tweaked, and "I don't know" was added as a possible response. The result was the NSF PIIR Cross-Disciplinary Toolbox, or the "STEM" instrument (for Science, Technology, Engineering, and Mathematics), as it was known within the Toolbox Project. (See Appendix A for the reformatted version of this Toolbox.)

The creation of a standardized STEM instrument reflected the Toolbox Project's goals in the analogue days. Because its research goals included understanding the impact of the Toolbox workshop in different contexts, standardization was necessary to facilitate comparison of results across workshops. Also, workshops were delivered almost exclusively to groups of STEM researchers, e.g., teams of collaborating scientists (the primary audience), students in graduate science courses, and leadership teams for research centers. The STEM instrument was a one-size-fits-all instrument for use by any group of scientists with a few hours on their hands and a desire for self-reflective philosophical conversation. Many groups of scientists from across the United States were introduced to the philosophical conversations of Toolbox workshops structured by the STEM instrument. TDI's research process at this point was exclusively analogue: its data were collected on paper forms on which participants had circled a rating score in response to each prompt. The data were copied into spreadsheets one score at a time, and record-keeping required filing stacks upon stacks of the paper instruments.

By 2010, the Toolbox Project had funding, a mission of improving scientific research through philosophically guided communication, a standardized instrument for conducting its mission, and a growing data set from Toolbox workshops. The next step was to publicize its activities. The team presented the project at the inaugural Science of Team Science (SciTS) Meeting in Chicago, which has since become one of the most important venues for TDI's work. The Toolbox Project also hosted its own meeting—*Enhancing Communication & Collaboration in Interdisciplinary Research*—in Coeur D'Alene, Idaho, in 2010. Michael Crow, President of Arizona State University, delivered the keynote address, and many leading thinkers about and practitioners of cross-disciplinarity attended. The content of the conference underscored the overwhelming importance of cross-disciplinary science to address the complex problems we face today.

As should be clear by this point, the history of TDI is a history of bringing philosophy to other disciplines. Influence has gone the other way as well, as TDI has brought insight and information from other disciplines into philosophy. The latter transfer was first discussed publicly in 2011 by O'Rourke and Crowley, which became the basis for two subsequent publications in philosophy journals (O'Rourke and Crowley 2013; Crowley et al. 2016). There is a lesson here for philosophers: engaged philosophy (see Chapter 6) is compatible with capital "P" Philosophy and is publishable in academic philosophy journals.

At the same time that the Toolbox Project was finding ways to root more deeply in academic philosophy, it was also expanding its mission beyond improving communication in cross-disciplinary science. In 2011, the Toolbox Project branched into translational medicine. The Toolbox team had been interested for several years in public and community health, but it wasn't until O'Rourke promoted the benefits of the Toolbox method for translational medicine at the Idaho State

University Division of Health Sciences Research Day that they found a health science partner. A collaboration ensued with an interested audience member, Lynn Schnapp, then at the University of Washington's Institute of Translational Health Science. This collaboration resulted in Schnapp et al. (2012), an article that also involved Troy E. Hall, a recent addition to the Toolbox team. In 2012, the Project went international, as O'Rourke and Crowley gave presentations at a conference on Integration in Biology and Biomedicine in Sydney, Australia. There they met Steven Orzack, who runs a non-profit research institute in the United States. His success at working outside of the university would encourage Toolbox to reimagine itself in the coming years.

3.4   THE ANALOGUE YEARS II: 2012–2015

In 2012, O'Rourke moved to Michigan State University (MSU), which became the new home for the Toolbox Project. MSU provided financial support, which paid for a post-doctoral research assistant, transcription of the audio recordings of the workshops, travel, and periodic retreats. The first retreat, held in 2014, gave Project members an opportunity to reflect on its mission and goals. With Orzack's input, which drew upon his experience with funding research at a non-profit research institute, the Toolbox Project began to consider new directions for its future.

From its inception, the Toolbox Project had operated both as a consultant to research groups and as a research group of its own. These activities generated some internal tension. The consulting activity fit several organizational models that would involve monetizing this activity. Some of these—for example, running the Project as a for-profit consulting business—were potentially at odds with maintaining the Project's academic research program. If the Project committed itself primarily to consulting and running workshops in order to raise money, it could reduce its research output, even though research had been key to improvement of its instruments. To continue this research, workshop fees would not be sufficient. It had financial support from MSU, but this support was not permanent. The Project could continue to seek grants to supplement its internal funding, but that would require a more substantial commitment by the team to grant writing. This, in turn, would take away from the capacity to run workshops, gather data, and help groups communicate better—that is, to fulfill the mission of the Toolbox Project.

Although the Project was beginning to ask itself fundamental existential questions, this did not immediately lead it to change course. It would still be housed in a university, have a dual mission providing consulting services and conducting research, and seek funding from external organizations. In 2013, it received an Ethics Education in Science and Engineering (EESE) grant from the NSF. This would take the Project in a new direction. The goal of the grant was to support ethical reasoning about the values and policy dimensions of environmental science through philosophical analysis and reflection. Instead of composing and implementing a one-size-fits-all instrument for the task, the Project designed a curriculum to help students write their own Toolbox prompts and run their own workshop. The motivation was to guide students in analyzing the ethical dimensions of their cross-disciplinary context and then reflect on those dimensions collectively in dialogue. This gave expanded

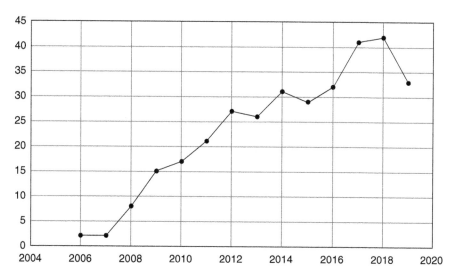

FIGURE 3.1 Graph of the number of Toolbox workshops delivered per year between 2006 and 2019.

pedagogical implementation of the Toolbox method, which, as noted above, had long been part of the Project.

Research productivity and workshop activity both continued at a steady pace during the Toolbox Project's first years at Michigan State. In 2014, the Project published a book entitled *Enhancing Communication and Collaboration in Interdisciplinary Research* (O'Rourke et al. 2014). Chapter 11 of that book detailed the mechanics of a Toolbox Workshop for the first time in print. The Project continued to extend its national and international reach, running workshops in San Juan, Puerto Rico, Stockholm, Sweden, and Waterloo, Canada. More workshops were run in 2014 than in any previous year (see Figure 3.1).

Even though the Project continued to be productive, it was failing to address the difficult existential questions of what it wanted to be and where it wanted to go. There is an important lesson here for any group that grows gradually from small-scale grassroots origins. Examining and refining one's institutional identity is hard work, especially within a team that has a pluralistic and democratic ethos. A team can continue to work well, even as it grapples with difficult, self-reflective questions, but at some point difficult existential decisions must be made.

3.5   THE DIGITAL YEARS: 2015–PRESENT

In the case of the Toolbox Project, the pressure to make these decisions mounted in 2015. That year, three MSU units announced that they would provide funding for the Toolbox Project. MSU's AgBioResearch, which was instrumental in bringing O'Rourke to MSU, announced that it would expand its support for the Project. The fit was good: AgBioResearch was eager to bring work in the humanities to research in agriculture and natural resources, and the Toolbox Project had shown its capacity to do this through its two NSF grants. Second, the College of Arts & Letters and the Provost's Office agreed to provide substantial funding for two years to help it transition into a more self-sustaining research effort. These units were thus eager to offer financial support to the Project, but they also wanted it to use this funding to find a way to sustain itself. These contributions provided the impetus for the Toolbox Project to transform itself into what it is today: the Toolbox Dialogue Initiative.

With this transition, TDI would move forward from its analogue past into its present-day digital reality. This involved reforms to its internal organization, its method of data collection, the way it runs workshops and engages with clients, and its outreach and marketing. Many of these changes involved literally digitizing Toolbox's practices, including shifting from the paper instruments to an online app. (See Chapter 11 for more on this.) Other changes were metaphorically digital, helping to make clear, defined, and discrete what had previously flowed without clear boundaries. The Toolbox Project had managed to function with loosely defined roles and responsibilities; TDI has clarified these, which has improved its performance. There is now a leadership team that comprises the Director (O'Rourke) and several associated Senior Members. This team guides a Data Manager, an Internal Communication Coordinator, a Social Media Coordinator, and Satellite Facilitators at several universities. An advisory board consists of long-standing allies, several of whom are from the community working on the Science of Team Science. Team meetings are scheduled routinely and have become more productive. These sorts of roles and practices are standard in well-functioning institutions; for TDI, they only came to be defined explicitly after almost a decade of its existence, as the team had successfully pursued its initiatives by assigning tasks on an as-needed basis.

The move from the Toolbox Project to TDI also involved an official change of name. The original name was problematic for several reasons. First, the terms "toolbox" and "toolkit" are not distinctive and are widely applied to interventions of all kinds, including those in philosophy and in the Science of Team Science. Second, "Toolbox Project" was already trademarked. For these reasons, the "Toolbox Dialogue Initiative" was selected because it kept the original word "Toolbox," added an important descriptor of its approach ("Dialogue"), and produced a distinctive acronym—"TDI." The lesson here: care when choosing a name is important!

With its new name, new organizational structure, new app for running workshops, and new methods for gathering and processing data, TDI has come a long way from its beginnings in Eigenbrode and O'Rourke's 2005 seminar. It has expanded beyond running Toolbox workshops for teams of scientists engaged in cross-disciplinary work and, for example, engaged with community stakeholders in Winnebago County, Illinois, and in East Africa (see Chapter 12). TDI's work in all of these instances involved the use of client-specific instruments, instead of the one-size-fits-all model of the analogue years. Some also involved post-workshop co-creation activities.

TDI also continues to seek ways to improve. In 2017, the Evaluation Center at Western Michigan University conducted an empirical evaluation of TDI's effectiveness. On the basis of TDI's own data and data they gathered from past participants, the Center determined that TDI's impact was positive (see Figure 3.2). For example, 90% of past participants who responded agreed that the prompts were effective

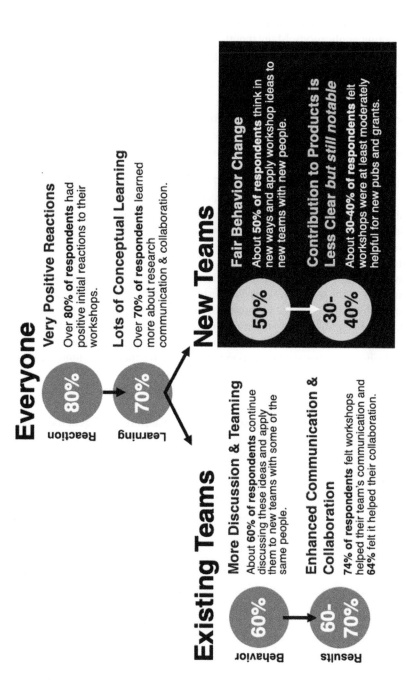

FIGURE 3.2 Summary of the findings from the Western Michigan University Evaluation Center's 2017 study of the impact of Toolbox workshops.

conversation starters and 77% felt more capable of collaborating with others from different disciplines. Approximately one-third of participants found their workshop at least moderately helpful in obtaining new funding and nearly half found it useful in co-authoring publications. Many reported wanting to continue to learn and practice new research dialogue skills; about 60% agreed they talked about these topics with their team and others after their workshop. This finding reminded TDI that it could do more to engage with participants after a workshop, possibly by creating an online community of practice, or offering follow-up training in collaborative skills.

At present, the mission of TDI is two-fold: to conduct academic research aimed at better understanding the nature of cross-disciplinary research, and to use what they learn to facilitate improvement in cross-disciplinary communication and collaboration. In keeping with this, TDI became a branch of Michigan State University's Center for Interdisciplinarity (C4I) in 2017. This connection provides TDI with the stability it needs to focus on projects for which it is a good fit. The stability also guarantees that TDI can employ the researchers it needs to manage and analyze its data. The lesson here is that sustainability of an endeavor like TDI can entail becoming part of a larger enterprise.

As the world's complexity increases, so do its problems. Organizations, methods, and team skills must also be complex in order to be up to the task of addressing these problems. TDI, along with other contributors to the Science of Team Science, can help in this effort by using "science to transform the ways researchers do science" (Hall 2017, p. 562). In addition to its empirical work on cross-disciplinary communication, TDI's engaged form of philosophy has a special role to play in this endeavor, just as it does anywhere that teams can function better through enhanced communication.

3.6   ACKNOWLEDGMENTS

Thanks to Stephen Crowley, Shannon Donovan, Sanford D. Eigenbrode, James Foster, Michael O'Rourke, Marisa A. Rinkus, and Stephanie E. Vasko for their input on this chapter. Thanks also to Todd Nagel for help with Figure 3.1.

3.7   LITERATURE CITED

Campbell, L. M. 2005: Overcoming obstacles to interdisciplinary research. Conservation Biology 19:574–577.

Crowley, S. J., C. Gonnerman, and M. O'Rourke. 2016: Cross-disciplinary research as a platform for philosophical research. Journal of the American Philosophical Association 2:344–363.

Eigenbrode, S. D., M. O'Rourke, J. D. Wulfhorst, D. M. Althoff, C. S. Goldberg, K. Merrill, W. Morse, M. Nielsen-Pincus, J. Stephens, L. Winowiecki, and N. A. Bosque-Pérez. 2007: Employing philosophical dialogue in collaborative science. BioScience 57:55–64.

Hall, K. L. 2017: What makes teams tick. Nature 551:562–563.

Heemskerk, M., K. Wilson, and M. Pavao-Zuckerman. 2003: Conceptual models as tools for communication across disciplines. Conservation Ecology 7:8.

Jakobsen, C. H., T. Hels, and W. J. McLaughlin. 2004: Barriers and facilitators to integration among scientists in transdisciplinary landscape analyses: A cross-country comparison. Forest Policy and Economics 6:15–31.

Kinzig, A. P. 2001: Bridging disciplinary divides to address environmental and intellectual challenges. Ecosystems 4:709–715.

Klein, J. T. 1996: Crossing Boundaries: Knowledge, Disciplinarities, and interdisciplinarities. University of Virginia Press, Charlottesville.

———. 2004: Interdisciplinarity and complexity: An evolving relationship. Emergence: Complexity and Organization 6:2–10.

O'Rourke, M., and S. J. Crowley. 2013: Philosophical intervention and cross-disciplinary science: The story of the Toolbox Project. Synthese 190:1937–1954.

O'Rourke, M., S. J. Crowley, S. D. Eigenbrode, and J. D. Wulfhorst. 2014: Enhancing Communication and Collaboration in Interdisciplinary Research. SAGE, Thousand Oaks, CA.

Schnapp, L. M., L. Rotschy, T. E. Hall, S. J. Crowley, and M. O'Rourke. 2012: How to talk to strangers: Facilitating knowledge sharing within translational health teams with the Toolbox dialogue method. Translational Behavioral Medicine 2:469–479.

Stokols, D., J. Fuqua, J. Gress, R. Harvey, K. Phillips, L. Baezconde-Garbanati, J. Unger, P. Palmer, M. Clark, S. Colby, G. Morgan, and W. Trochim. 2003: Evaluating transdisciplinary science. Nicotine & Tobacco Research 5:21–39.

Taper, M. L., and S. R. Lele. 2004: The Nature of Scientific Evidence: Empirical, Statistical, and Philosophical Considerations. University of Chicago Press, Chicago.

4 The Landscape of Challenges for Cross-Disciplinary Activity

Shannon Donovan

4.1 INTRODUCTION

In his article entitled "Five Arguments Against Interdisciplinary Studies," Benson (1982) argues that while much effort is spent praising the merits of interdisciplinary studies, more attention should be paid to the pitfalls and challenges of this type of integrative work and curriculum. He discusses why interdisciplinary study is inappropriate in an undergraduate classroom, claiming that (p. 46) "the student is exposed to a semester-long variety show, doubtless interesting, but of very little long-term educational value" and that (p. 47) "integrative studies faculty are, for the most part, second-class scholars, exiles and refugees from the disciplinary departments, where they either failed to measure up or found themselves incapable of sustaining the kind of rigor and focus required for success in disciplinary scholarship."

Benson's strong opinions do not reflect the current state of interdisciplinary education, but they are nevertheless useful. He is correct to note that interdisciplinary scholarship has many challenges, and those challenges extend beyond the undergraduate classroom and tax graduate students and researchers alike. As their educational focus tends to narrow, graduate students struggle to find the interdisciplinary mentorship they need to address the complex problems of the real world (Rhoten and Parker 2004). Similarly, scholars face diverse and complex obstacles when attempting interdisciplinary and, more generally, cross-disciplinary research. Examples of such obstacles include institutional barriers (Rhoten and Parker 2004), the domain-specific structure of science (MacLeod 2018), the historical lack of well-respected, cross-disciplinary journals (Campbell 2005), miscommunication due to disciplinary assumptions (Donovan et al. 2015), and ideological and/or philosophical differences between interdisciplinary collaborators (Campbell 2005). Nevertheless, there is no shortage of researchers willing to engage in cross-disciplinary research because many of today's most pressing questions are complex and cross-disciplinary in nature and require an integrated approach to problem solving. Moreover, the frequency of collaborative work has increased due in part to the rise of social media and online networking opportunities such as Google Scholar and ResearchGate (Van Noorden 2015) and the increased ability of experts to work together from around the globe using virtual platforms. This chapter explores some of the challenges faced by cross-disciplinary scholars and the questions they should consider before and during their work. Cross-disciplinary work of this sort is central to TDI, which has helped hundreds of teams identify and examine the range of cross-disciplinary challenges they encounter.

4.2   COLLABORATION: A DIFFICULT ART AND SCIENCE

Collaboration is both the art and science of working together to achieve a common goal. Working in any context can be a challenge. Barriers to collaboration include lack of trust (such as not wanting to share information and fear of failure), differences in work ethic (such as meeting deadlines, following through with commitments, and accomplishing tasks), differences in communicative style (such as email etiquette, frequency, and rate of response to emails and phone calls, and how one participates in conference calls), failure to set clear expectations about roles and boundaries, and interpersonal differences (Wondolleck and Yaffee 2000; Cheruvelil et al. 2014).

These barriers appear in the context of research collaboration as well. For example, collaboration within a discipline with colleagues who share much in common can be difficult due to lack of trust and differences in work ethic and communicative style. Barriers like these can explain why an individual who does great work need not be a great or even a good collaborator. Collaboration across disciplines adds additional complexity to this already complicated process. Cross-disciplinary work is innately challenging, but one dimension of difficulty that is especially important for TDI concerns the complex epistemic issues that arise when diverse perspectives meet.

4.3   CHALLENGES OF COLLABORATING ACROSS THE DISCIPLINES

Challenges facing cross-disciplinary researchers are well documented (e.g., Brewer 1999; Campbell 2005; Aagaard-Hansen 2007; Morse et al. 2007; Willyard 2013). Recognizing and addressing these challenges early in the research process can help teams be successful.

4.3.1   GETTING OUT OF THE BOX

When working on complex problems, disciplinarity can box you in by narrowing your focus and restricting your methodological options; cross-disciplinarity, by contrast, can help you get out of the box and meet the complexity of the problem with complexity in response. How do you know if a project is *truly* cross-disciplinary, and how are interdisciplinary, multidisciplinary, transdisciplinary approaches different? As noted in Chapter 1, these categories are part of a spectrum (Morse et al. 2007; Hall et al. 2012). During TDI workshops, participants often struggle with these categories and the associated language. For example, one team member described their team as multidisciplinary because they work independently to reach their goals while another member considered them to be interdisciplinary because they were working towards a common, integrated goal. However, the challenge of getting out of the box is not about names; rather, it is about agreeing on the best type of cross-disciplinarity to pursue based on project and individual researcher needs. Regardless of labels, to be truly cross-disciplinary, a project needs to combine diverse disciplinary inputs that are determined by project needs and proposed outcomes.

4.3.2 THE TASK OF LETTING GO

One of the fundamental challenges in cross-disciplinary research comes from the need to let go. The task of letting go of being an expert in all areas of a cross-disciplinary project, giving up some of the control of a project, and trusting that others have the same level of expertise as you (albeit from different disciplines) is a challenge in and of itself (Mesmer-Magnus and DeChurch 2009). But, as Benson (1982) suggests, it is difficult to master one discipline and it is foolish to try to master multiple disciplines if the goal is to generate a cross-disciplinary response and it is possible to collaborate. Given this, true cross-disciplinary research—i.e., cross-disciplinary research that combines expertise from different disciplines—typically creates a need to rely on the knowledge and skills of others. Over the years, TDI members have learned that this can be especially difficult for (1) those who have limited experience learning from others outside of their disciplines, (2) those who have limited experience participating in group processes, (3) those who are not very interested in understanding new or different research approaches, or (4) those who have primarily focused on basic research questions or are not very interested in applying the results of their work. Any of these factors can decrease practitioners' willingness or readiness to learn from and rely upon their team members, which in some cases can undermine a team's success.

4.3.3 FALSE STARTS

Any cross-disciplinary research project should begin with a clear objective, a refined research question, and clear understanding of and agreement on project goals. Researchers should then confirm they have the expertise needed to reach their goals. Once the team is formed, a project plan should be established and modified to meet the needs of the individuals. Specific aspects of a project plan will vary depending on research needs and project personnel, but plans could include the following elements: to what degree individual research sub-questions need to align, the appropriateness of individual researchers' scope, anticipated timelines for generating and analyzing individual data, and a framework for integrating individual findings. Making such a plan will help team members recognize their gaps, be transparent about their limitations, and create opportunities for changing course when needed. Recognizing gaps early on can help team members understand to what degree they are prepared to engage in a cross-disciplinary activity and assess what changes they need to make (e.g., adjust their research question, change their proposed outcomes, modify their scope, recruit additional members to their team). However, cross-disciplinary activity can also happen in more of a choose-your-own-adventure model where the activities start before the project foundation is fully formed. Indeed, much of TDI's own history can be understood as proceeding in this way (see Chapter 3). In some cases, this model can involve backpedaling on antecedently conducted disciplinary activity in an attempt to create cross-disciplinarity post hoc.

However, backpedaling into cross-disciplinarity can cause a number of problems. Under those circumstances, collaboration may amount to a hodge-podge of disciplinary activity meshed together, which may not result in the desired outcome. An example of this was seen by one team of graduate students who did not have proper mentorship on how to develop and conduct cross-disciplinary research. Each team member worked independently to address their own research questions and then tried to weave their disciplinary research into a cross-disciplinary product. The result was a tangled mess and a false start that required time and money to undo and did not end up addressing a clear cross-disciplinary question. Laying out cross-disciplinary plans and defining cross-disciplinary research questions can be difficult (Hughes 2007), but starting with a clear problem and set of questions that you are trying to address makes an already complex process easier.

4.3.4 TOMATO, TOMAHTO: COMMUNICATING EFFECTIVELY

Working with others is hard, especially when they speak different disciplinary languages. To function effectively, teams must communicate effectively; in fact, without effective communication, cross-disciplinary teams may just quit. Effective communication—and so, potentially, persevering within a project—requires developing a shared understanding of the different terms and concepts used by contributing disciplines (Bracken and Oughton 2006; Eigenbrode et al. 2007; Thompson 2009; Hall et al. 2012; Donovan et al. 2015). Overcoming language differences does not require cross-disciplinary practitioners to be fluent in other disciplinary languages; rather, they should strive to (1) accept that other interpretations and understandings exist, and (2) gain a basic understanding of concepts and approaches that are "foreign" to them (Mesmer-Magnus and DeChurch 2009; Klein 2010). Furthermore, cross-disciplinary practitioners should consider integrative activity as a chance to share the culture of their disciplines with one another. This creates opportunities for localized work where scholars can bring their unique perspectives and experiences to the table (Crowley et al. 2010)

A basic understanding of these concepts and approaches is valuable because it can support the nuanced differences of technical terms used across disciplines, which can be crucial to the success of cross-disciplinary projects (Wear 1999; Marzano et al. 2006; Morse et al. 2007; Thompson 2009). For example, for one team member, the word "significance" may refer to a statistical concept, while for another it may concern the importance of a research theme or concept that figures in the project. Not having a clear understanding of such terminological practices could result in miscommunications that derail the cross-disciplinary research process.

To better understand the consequences of such language differences, Donovan et al. (2015) analyzed responses to these prompts in Toolbox sessions:

Scientific research (applied or basic) must be hypothesis driven.
There are strict requirements for determining when empirical data confirm a tested hypothesis.

The findings show that the term "hypothesis" is used in a multitude of ways, including as the driver of research study, as an assumption about a topic, as the criterion for rejecting an assumption, or as an initial question for an exploratory process. Such differences in interpretation can impede collaboration if they go unacknowledged by project collaborators.

4.3.5 Your Scale Is Out of My Scope

Researchers may differ in how they view the scale (or extent) and the scope (or breadth) of a project because of their different disciplinary training. For example, a microbiologist is likely to tackle a project from a different scale and with a different scope than a landscape ecologist, even if they are conducting their work within the same geographic area, while a social scientist can study a smaller, specific population or a larger, more general population depending on their research question. Different scales and scopes need not be deal breakers, but they do need to be brought to light early in order to mitigate integration problems arising from cross-disciplinary differences later in a project (O'Rourke et al. 2016).

For example, during a Toolbox session, a team discussed struggles over the alignment of the scale of methods they used to collect soil and vegetation samples. These struggles led team members to adjust their methodology so that they worked within the same plots, and they developed a research design that would allow the soil scientist on the team to extrapolate findings to larger areas. Modifying their research questions and goals allowed the team to adjust their scale and scope, resulting in truly cross-disciplinary work. Without such modifications, researchers may find at the end of their project that their results are not integrated and they do not reap the full benefits of the cross-disciplinary process.

4.3.6 The Uneven Playing Field

Power struggles that hinder team effectiveness can be found in all kinds of collaborations, including cross-disciplinary projects. For example, Van Bunderen et al. (2018) describe how teams that have a hierarchical structure can promote more individual-level concerns and can hinder team function, particularly concerning resource availability. Furthermore, in groups where power is unevenly distributed between team members, conflict resolution within the team tends to be more difficult (Jones 2012).

Members of TDI have seen many such struggles. For example, one Toolbox session revealed a lack of graduate student involvement in the project's development because project leadership marginalized students and post-docs. In another instance, junior faculty spoke little, and when they did, senior faculty members often interrupted. In yet another session, the facilitator observed that all the women sat on one side of the table and all the men sat on the other, creating a visual, and perhaps actual, imbalance of power. Such power dynamics can be found anywhere and are not reflective of any one discipline. It is important to note that when working with others in cross-disciplinary projects, an imbalance of power may arise that has the potential to create barriers to effective collaboration.

4.3.7   INTEGRATION AND THE ROAD LESS TRAVELED

Little work has been done to understand how cross-disciplinary researchers integrate their disciplinary inputs (O'Rourke et al. 2016). There are no widely established practices for how to do so at this time. While the theories, literature, and methods used for integrating and analyzing disciplinary inputs will vary across teams, we argue that meaningful integration depends on the development of a clearly defined research question. According to Cosens et al. (2011, pp. 123–124).) formulation of a question "appears simple until integration across multiple disciplines is considered." While a difficult process, the act of developing a cross-disciplinary question helps reveal a team's initial barriers to integration (Cosens et al. 2011).

The research question serves as the basis for the subsequent research plan in which a team identifies the various scales, scopes, data generation techniques, data analysis methods, and disciplinary differences, among other things; this plan will aid team members as they work to integrate their project. Graduate-level courses, such as the one described by Larson et al. (2011), and interdisciplinary research guides, such as those offered by the University of Iowa Office of the Vice President for Research (2017) and Lyall and Meagher (2007), offer advice on how to develop such a plan. These can help reduce the amount of time needed for cross-disciplinary research, which often takes more time than research conducted by individuals (Rhoten and Parker 2004; Lee 2006; Else 2016).

4.3.8   TRAINING CROSS-DISCIPLINARIANS

There is little training available for educators on how to teach students to do cross-disciplinary research (Hall et al. 2012), despite the fact that cross-disciplinary scholarship will be necessary to address many pressing societal problems (National Academy of Sciences 2005). Such training requires commitment by instructors and by institutions (Styron 2013). Most faculty do not know how to teach such a course. Even with the existence of cross-disciplinary graduate programs (e.g., University of California San Diego Training Center, Institute for Cross-Disciplinary Engagement at Dartmouth), models for conducting cross-disciplinary research are rare. Graduate students often receive little mentorship from faculty. Graduate students may also be interested in pursuing cross-disciplinary research but be limited by their advisor (Rhoten and Parker 2004).

Most faculty with cross-disciplinary expertise have gained it through experience, rather than through formal training in cross-disciplinary research. They may have limited experience teaching outside of their discipline, which may make it more difficult to convey experiences with cross-disciplinary research to students. Faculty who are interested in training students to conduct cross-disciplinary research must be committed to teaching outside of their disciplinary comfort zone and dedicated to transforming their own ways of knowing (Rhoten and Parker 2004; Hall et al. 2012).

Graduate students can feel a contradiction when engaging in cross-disciplinary activity. When so engaged, students gain a great breadth of skills, yet most of them are also taught that their focus should narrow as they continue their education. This contradiction can be difficult for some students to manage, particularly if their advisors are not engaged in cross-disciplinary activities. The difficulty of managing this contradiction may be especially challenging when graduate students need to focus on qualifying exams and writing their theses (Rhoten and Parker 2004).

While there have been a few efforts to assess how students perform when conducting cross-disciplinary research (e.g., Hackett and Rhoten 2009), there is little guidance for instructors who wish to evaluate the role a student plays in making the cross-disciplinary whole bigger than the sum of the individual parts. The lack of instructor guidance is due in large part to the extreme difficulty of assessing a student's contribution to a team effort. This is especially true when the stakes are high, such as when the team effort involves production of theses or dissertations (Rhoten and Parker 2004; Bosque-Pérez et al. 2016). TDI helps call attention to these gaps by presenting opportunities to engage in philosophical discussions that highlight the disciplinary assumptions that may reduce the effectiveness of a cross-disciplinary research team.

4.3.9 OTHER CROSS-DISCIPLINARY CHALLENGES FACING THE PROFESSIONAL

Academia, government, and other organizations often do not have the infrastructure to evaluate and support cross-disciplinary work (Hall et al. 2018). These institutions may approve of such work but not promote the cross-disciplinary process, which requires extra time both for effective collaboration and for management of the logistical difficulties encountered when working across disciplines and institutions.

There is often an underlying fear that engaging in cross-disciplinary work will hinder future opportunities because employers or supervisors will not appreciate such work (Rhoten and Parker 2004). For example, because cross-disciplinary research may be valued at one institution but not at another, faculty members who are committed to engaging in cross-disciplinary work may worry such activity will work against them should they apply for a job at a different university; the same is true in the other direction for those who are not committed to cross-disciplinarity, although there are fewer institutions that will expect cross-disciplinary fluency (Rhoten and Parker 2004). Furthermore, cross-disciplinary research often takes longer than disciplinary research, which can result in cross-disciplinary practitioners appearing to be less productive than their disciplinary-focused peers (Leahey et al. 2017).

When undertaking a cross-disciplinary research agenda, a practitioner should consider these questions:

Is cross-disciplinary research recognized and rewarded at my institution?

What is the reward for participating in cross-disciplinary research?

If cross-disciplinary research is recognized and rewarded, then a practitioner should feel confident about moving forward. If not, then a practitioner should consider their options, one being to assess what is needed to in order to create recognition and reward for such research.

4.4 CHALLENGING CROSS-DISCIPLINARY QUESTIONS

In order to avoid the cross-disciplinary challenges outlined above, a practitioner should answer these questions:

Does your cross-disciplinary research project have a purpose and specific research questions? If not, the research should begin by collectively defining these. A specific research question and the group of researchers appropriate for addressing this question are the foundation for any successful cross-disciplinary research project.

What scale and scope are appropriate for the project, given the purpose and research question? Are compromises among collaborator perspectives on scale and scope required to achieve the goals of the project? These compromises could involve adjustment to the number of samples, the locations of data collection, and the methodologies used to ease data integration later in the research process.

Who needs to be part of the research process? Given your research question, assess whether the team has the needed expertise. An early assessment of what expertise is needed to answer your research question enhances the probability that the results will be truly cross-disciplinary.

Are your team members committed to using a cross-disciplinary approach? Determine the resources team members are able to contribute (e.g., money, time, personnel).

What are the desired products? Determine how data sets will come together and the mechanisms you will use for cross-disciplinary data integration and analysis.

4.5   CONCLUSION

Given the challenges of cross-disciplinary research, why do it? Here are a few reasons to move forward:

Cross-disciplinary research is necessary for addressing our most complex problems. Such research is high risk/high reward, but the likelihood of high reward is improved when the challenges are effectively negotiated.

Cross-disciplinary research gets easier with time. Practitioners become better at recognizing the importance of questions that cross disciplinary lines, better at hearing responses, and better at adjusting approaches so as to enhance cross-disciplinary collaboration.

Although it is impossible to create a one-size-fits-all approach, lessons learned from past cross-disciplinary activity can often be applied to new projects. Such lessons include talking about differences in philosophy and in disciplinary approaches at the beginning of a project, including the necessary disciplines from the beginning of the project, and developing protocols for how the team will communicate.

While cross-disciplinary research will be challenging, a practitioner's comfort with the process should increase with experience. Individuals participating in cross-disciplinary activities will learn the range of skills and knowledge they can contribute, the elements that create a high-functioning team, and the mechanisms used to communicate across disciplines. Prior experiences help practitioners evaluate whether or not a cross-disciplinary activity will provide a reward that is worth the challenge.

If you are willing to risk navigating the landscape of challenges to tackle complex problems, then break free of your disciplinary silos and collaborate!

4.6 ACKNOWLEDGMENTS

I thank Toolbox workshop participants for helping me understand the diverse challenges facing cross-disciplinary practitioners. Special thanks go to the University of Idaho Integrative Graduate Education and Research Traineeship (IGERT) faculty and fellows. I also thank Graham Hubbs, Steven Orzack, and Michael O'Rourke for their editing assistance.

4.7 LITERATURE CITED

Aagaard-Hansen, J. 2007: The challenges of cross-disciplinary research. Social Epistemology 21:425–438.

Benson, T. L. 1982: Five arguments against interdisciplinary studies. Issues in Interdisciplinary Studies 1:38–48.

Bosque-Pérez, N. A., P. Z. Klos, J. E. Force, L. P. Waits, K. Cleary, P. Rhoades, S. M. Galbraith, A. L. B. Brymer, M. O'Rourke, S. D. Eigenbrode, B. Finegan, J. D. Wulfhorst, N. Sibelet, and J. D. Holbrook. 2016: A pedagogical model for team-based, problem-focused interdisciplinary doctoral education. BioScience 66:477–488.

Bracken, L. J., and E. A. Oughton. 2006: "What do you mean?" The importance of language in developing interdisciplinary research. Transactions of the Institute of British Geographers 31:371–382.

Brewer, G. D. 1999: The challenges of interdisciplinarity. Policy Sciences 32:327–337.

Campbell, L. M. 2005: Overcoming obstacles to interdisciplinary research. Conservation Biology 19:574–577.

Cheruvelil, K. S., P. A. Soranno, K. C. Weathers, P. C. Hanson, S. J. Goring, C. T. Filstrup, and E. K. Read. 2014: Creating and maintaining high-performing collaborative research teams: The importance of diversity and interpersonal skills. Frontiers in Ecology and the Environment 12:31–38.

Cosens, B., F. Fiedler, J. Boll, L. Higgins, B. K. Johnson, E. Strand, P. Wilson, M. Laflin, R. Szostak, and A. Repko. 2011: Interdisciplinary methods in water resources. Issues in Interdisciplinary Studies 29:118–143.

Crowley, S. J., S. D. Eigenbrode, M. O'Rourke, and J. D. Wulfhorst. 2010: Cross-disciplinary localization: A philosophical approach. MultiLingual 114:1–4.

Donovan, S. M., M. O'Rourke, and C. Looney. 2015: Your hypothesis or mine? Terminological and conceptual variation across disciplines. SAGE Open 5:1–13.

Eigenbrode, S. D., M. O'Rourke, J. D. Wulfhorst, D. M. Althoff, C. S. Goldberg, K. Merrill, W. Morse, M. Nielsen-Pincus, J. Stephens, L. Winowiecki, and N. A. Bosque-Pérez. 2007: Employing philosophical dialogue in collaborative science. BioScience 57:55–64.

Else, H. 2016: 4 Years, 7 Months to Publish Article. Inside Higher Education. Downloaded from www.insidehighered.com/news/2016/12/02/scholar-complains-how-long-it-can-take-publish-interdisciplinary-science.

Hackett, E. J., and D. R. Rhoten. 2009: The Snowbird Charrette: Integrative interdisciplinary collaboration in environmental research design. Minerva 47:407–440.

Hall, K. L., A. L. Vogel, B. A. Stipelman, D. Stokols, G. Morgan, and S. Gehlert. 2012: A four-phase model of transdisciplinary team-based research: Goals, team processes, and strategies. Translational Behavioral Medicine 2:415–430.

Hall, K. L., A. L. Vogel, G. C. Huang, K. J. Serrano, E. L. Rice, S. P. Tsakraklides, and S. M. Fiore. 2018: The science of team science: A review of the empirical evidence and research gaps on collaboration in science. American Psychologist 73:532–548.

Hughes, J. W. 2007: Environmental Problem Solving: A How-to Guide. University Press of New England, Lebanon, NH.

Jones, R. M. 2012: Introducing the elephant in the room: Power. Public Administration Review 72:417–418.

Klein, J. T. 2010: A taxonomy of interdisciplinarity. Pp. 15–30 *in* R. Frodeman, J. T. Klein, and C. Mitcham, eds. The Oxford Handbook of Interdisciplinarity. Oxford University Press, Oxford.

Larson, E. L., T. F. Landers, and M. D. Begg. 2011: Building interdisciplinary research models: A didactic course to prepare interdisciplinary scholars and faculty. Clinical and Translational Science 4:38–41.

Leahey, E., C. M. Beckman, and T. L. Stanko. 2017: Prominent but less productive: The impact of interdisciplinarity on scientists' research. Administrative Science Quarterly 62:105–139.

Lee, C. 2006: *Perspective: Peer review of interdisciplinary scientific papers.* Nature Peer to Peer. Downloaded from http://blogs.nature.com/peer-to-peer/2006/06/perspective_peer_review_of_int.html.

Lyall, C., and L. Meagher. 2007: A Short Guide to Building and Managing Interdisciplinary Research Teams. The University of Edinburgh, Edinburgh.

MacLeod, M. 2018: What makes interdisciplinarity difficult? Some consequences of domain specificity in interdisciplinary practice. Synthese 195:697–720.

Marzano, M., D. N. Carss, and S. Bell. 2006: Working to make interdisciplinarity work: Investing in communication and interpersonal relationships. Journal of Agricultural Economics 57:185–197.

Mesmer-Magnus, J. R., and L. A. DeChurch. 2009: Information sharing and team performance: A meta-analysis. Journal of Applied Psychology 94:535–546.

Morse, W. C., M. Nielsen-Pincus, J. E. Force, and J. D. Wulfhorst. 2007: Bridges and barriers to developing and conducting interdisciplinary graduate-student team research. Ecology and Society 12:1–14.

National Academy of Sciences. 2005: Facilitating Interdisciplinary Research. National Academies Press (US), Washington DC.

O'Rourke, M., S. J. Crowley, and C. Gonnerman. 2016: On the nature of cross-disciplinary integration: A philosophical framework. Studies in History and Philosophy of Science Part C: Studies in History and Philosophy of Biological and Biomedical Sciences 56:62–70.

Rhoten, D., and A. Parker. 2004: Risks and rewards of an interdisciplinary research path. Science 306:2046.

Styron Jr., R. A. 2013: Interdisciplinary education: A reflection of the real world. Journal of Systemics, Cybernetics and Informatics 11:47–52.

Thompson, J. L. 2009: Building collective communication competence in interdisciplinary research teams. Journal of Applied Communication Research 37:278–297.

University of Iowa Office of the Vice President for Research. 2017: Best Practices for Developing Interdisciplinary Research Teams.

Van Bunderen, L., L. L. Greer, and D. Van Knippenberg. 2018: When interteam conflict spirals into intrateam power struggles: The pivotal role of team power structures. Academy of Management Journal 61:1100–1130.

Van Noorden, R. 2015: Interdisciplinary research by the numbers. Nature 525:306–307.

Wear, D. N. 1999: Challenges to interdisciplinary discourse. Ecosystems 2:299–301.

Willyard, C. 2013: *The agony and ecstasy of cross-disciplinary collaboration.* Science. Downloaded from www.sciencemag.org/careers/2013/08/agony-and-ecstasy-cross-disciplinary-collaboration.

Wondolleck, J. M., and S. L. Yaffee. 2000: Making Collaboration Work: Lessons from Innovation in Natural Resource Management. Island Press, Washington DC.

5 Communication and Integration in Cross-Disciplinary Activity

Michael O'Rourke
Brian Robinson

5.1 INTRODUCTION

The previous chapter describes the hurdles that must be cleared by cross-disciplinary collaborators if their projects are to succeed. Principal among these obstacles are those related to integration and communication. Success in collaborative research and practice requires the *integration* of different perspectives, which entails that the perspectives influence the final outcome and do not get lost. To accomplish this kind of integration, there must be *communication* among collaborators. Both integration and communication require cooperation among collaborators, where this typically involves both social interaction and epistemic exchange. In this chapter, we expand on these observations and provide an analysis of communication, integration, and the relationship between them. The following questions guide our work: What are communication and integration? What special challenges do communication and integration pose? and How can these challenges be overcome?

5.2 COMMUNICATION

At the heart of interdisciplinarity is communication—the conversations, connections, and combinations that bring new insights to virtually every kind of scientist and engineer. … Without sustained and intense discussion … and without special effort by researchers to learn the languages and cultures of participants in different traditions, the potential interdisciplinary research might not be realized and might have no lasting effect.

National Academy of Sciences 2005, pp. 19–21

This statement emphasizes the centrality of communication to research that crosses disciplines. The mention of "conversations, connections, and combinations" underscores the importance of human relationships and interactions to epistemic innovation and emphasizes the idea that epistemic gains are social products rather than individual achievements. Communication is crucial for collaboration but especially so for *cross-disciplinary* collaboration, given the need to combine different epistemologies and make connections across different literatures and knowledge bases. The "intense" discussions and learning that are involved in successful cross-disciplinary collaboration require each collaborator to function as both teacher and learner.

This intensity is exemplified by exchanges within TDI, which itself is a cross-disciplinary research project. The Toolbox approach originated in a graduate seminar in which it became clear that philosophy could help cross-disciplinary collaborators focus on implicit assumptions that framed their scientific practice (Bosque-Pérez et al. 2016; Piso et al. 2016). The philosophers in the seminar suggested that the philosophical distinction between *epistemology* and *metaphysics* could be used as a framework for mapping conceptual schemes used by different disciplinary scientists. This suggestion was met with dismay—if philosophy is going to help, surely *this* sort of top-down philosophy, loaded with technical terminology, was the wrong tool for the job. Eventually, the group agreed to build a tool around this distinction, but only after intense conversations in which the philosophers explained how valuable it could be as an organizing principle.

This example illustrates the importance of communication for organizing collab-orative practice. All members of the seminar expected the philosophers to take the lead in identifying the philosophy that would inform the emerging approach, but they differed in regard to what form that philosophy would take. The philosophers had to connect what others took philosophy to be and what they had proposed. By doing so, they created a common conception that served to anchor consensus-based development of the Toolbox approach. It also illustrates the damage that can be done by the use of mysterious terms (e.g., "epistemology") or the use of the same term with different meanings (e.g., "philosophy"). Additional communication challenges include the use of different terms with the same meaning, different communication styles, and different ways of relating to conversational partners.

In what follows, we begin by identifying the mode of research communication that primarily interests us, namely communication among research collaborators. We then survey the communication literature to support a distinction between *infor-mational* and *sociopragmatic* communication, each of which foregrounds important aspects of communication among scientists. Finally, we turn to communication in cross-disciplinary contexts, highlighting the ways in which it can be especially chal-lenging, including the need to learn from others and the need to defer in situations where one's own expertise is not relevant.

5.2.1 The Direction of Research Communication

Much of the work that TDI has done to enhance communicative capacity within cross-disciplinary research teams has involved scientific research teams. Communication is often a point of emphasis in such teams, especially when they are larger, more diverse, and engaged in more controversial research. For example, Woody Weeds, a large project that focuses on developing strategic land management policies for woody invasive plants in Ethiopia, Kenya, and Tanzania, has partnered with TDI to address communication issues across a complex set of disciplines, work packages, and stakeholder groups. (Woody Weeds is discussed at greater length in Chapter 11.)

Two other large projects that partnered with TDI, the Regional Approaches to Climate Change (REACCH) and the Bio-computational Evolution in Action Consortium (BEACON), had communication-focused sessions at their launch meetings. TDI ran workshops in both meetings that were designed to help the participants identify potential obstacles to cross-disciplinary collaboration, but both meetings also had sessions focused on how to communicate with various stakeholder groups, the press, and the public in ways that avoided unhelpful creation of controversy around climate change for REACCH and around evolution for BEACON.

REACCH and BEACON both recognized the importance of the audience to science communication and designed sessions to help their participants understand the audience with whom they were communicating. We agree with this emphasis, and in this chapter we deploy the audience as a frame for thinking about communication. With the relationship between speaker and audience in mind, we distinguish between communication *in*, communication *across*, and communication *out*. Communication *in* concerns interactions with collaborators. Communication *across* concerns communication with colleagues who are not collaborators. Communication *out* concerns communication with the public, policymakers, and the press. (See Figure 5.1.)

Much of the literature on research communication focuses on communication *out*, and, specifically, on communication of science to non-scientists. This sort of communication, specifically as it is aimed at policymakers and politicians, is crucial because it can help sustain and build public funding of science (Brod 2014). For scientific research on politically controversial topics, such as research on evolution or climate change, effective science communication maneuvers around fruitless debates that could undermine public support for scientific research and

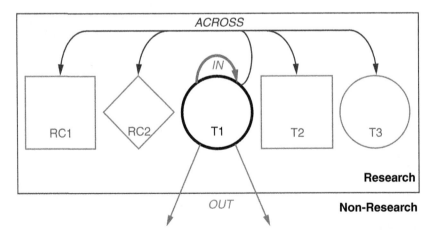

FIGURE 5.1 Directions of communication involving research team T1. *In* concerns communication within T1, *across* concerns communication with other researchers and research communities (e.g., research communities RC1 and RC2, as well as other teams, such as T2 and T3—the shapes signify differences in the structure of the communities), and communication *out* concerns communication with various non-research groups.

education (Stocklmayer 2001). Successful communication of scientific research on climate change could be the difference between life and death for many species, including our own (cf. McCright and Dunlap 2011; McCright et al. 2013). There is a growing field of science communication that focuses on communication *out*, including work that emphasizes philosophical challenges associated with this type of research, such as the need to understand and be competent with respect to values (Dietz 2013) and the importance of appreciating differences in epistemology (Suldovsky et al. 2018).

Communication *across* creates and maintains the connective tissue within and between research communities. Such communities include those that are less cohesive, such as universities, and those that are more cohesive, such as disciplines. Scientists must communicate with their colleagues because their professional success and the success of science as an enterprise depend on it. Unlike communication *out*, communication *across* is acknowledged as a professional research requirement. Disciplinary institutionalization involves the creation of opportunities to present and respond to research results. Blogs, conferences, workshops, and meetings of professional societies enable researchers to give and receive input from their colleagues on works in progress. Newsletters, journals, edited volumes, and books offer opportunities to contribute more polished products to the ongoing conversations that give form to a discipline.

Further, these venues for communication *across* contribute to professionalization in a discipline, both for those who contribute their research products but also for those who listen and comment. Conference commentators and audience members, as well as peer reviewers, interpret the works of their colleagues and offer critical feedback that is (typically) intended to help improve the quality of those works. When a person is trained in a discipline, identifying what is important in a professional presentation or manuscript and offering useful commentary are two abilities that one is expected to develop. Participation in research commentary and review is a key vehicle for creating the sense of disciplinary identity that is for many an important part of what it is to be a professional researcher.

Our primary concern in this chapter, though, is communication *in*. As a research collective, TDI produces research on the communicative dynamics and relationships that enable a group of researchers to coalesce as a team. There is no shortage of literature on team dynamics (e.g., Falk-Krzesinski et al. 2011; Fiore et al. 2010; Keyton and Beck 2010; Kozlowski 2015; National Research Council 2015). However, the focus of TDI is communication *in* among heterogeneous collaborators in cross-disciplinary settings that enables them to interact effectively. Communication among cross-disciplinary collaborators is communication *in*, but it bears important relationships with the other types of communication described above. Multi- and interdisciplinary communication among teammates requires the exchange of information across disciplines, which can seem like communication *across*, in that it involves communication with people who are likely unfamiliar with the epistemology or ontology of one's own discipline (Eigenbrode et al. 2007). Collaborative transdisciplinary research, understood to involve non-academic stakeholders, combines the disciplinary complexity of interdisciplinary research with the need to communicate with partners who may not be familiar with the practices and standards of academic research. This resembles communication *out* to stakeholders who are not part of one's team. Understanding how communication *in* relates to the other modes of communication is critical to a proper appreciation of collaborative, cross-disciplinary research, and these relationships also highlight the special communication challenges associated with this mode of research practice.

5.2.2 WHAT IS COMMUNICATION?

This question has a multitude of answers; indeed, an entire discipline, communication science, is devoted to making sense out of communication. Surveying these answers would necessitate a book in itself. Instead, we present a way of thinking about communication, grounded in both communication science and philosophy, that highlights the aspects of central importance to TDI. We begin by providing a way of thinking about communication that TDI endorses. Focusing on an example of communication, we provide an answer by using a model that comprises two dimensions, the *informational* dimension and the *sociopragmatic* dimension. We develop this model by crossing a pair of philosophical distinctions, the first a distinction between the *agential* and the *linguistic* elements of the example, and the second a distinction between the *content* and *form* of the example. We conclude this section by situating our model in the communication literature.

Communication is, first and foremost, cooperative, wherein "each participant recognizes a common purpose or set of purposes, or at least a mutually accepted direction," and then everyone makes their "conversational contribution such as is required at the stage at which it occurs, by the accepted purpose or direction of the talk exchange in which [they] are engaged" (Grice 1989, p. 26). Furthermore, it distinguishes the give and take of collaboration from the research equivalent of toddler "parallel play," which would involve researchers working in close proximity on their own projects without interaction or exchange (cf. Keyton et al. 2008). Communication underwrites meaningful interaction and exchange—it comprises the "conversations, connections, and combinations" that glue the team together. This points in the direction of the conception of communication that we favor: communication is the *co-construction of meaning in pursuit of a goal* (Keyton and Beck 2010; Hall and O'Rourke 2014). This characterization emphasizes the collaborative construction of meaning for the purpose of accomplishing some end. The purpose provides a constraint on the process and a way of determining if it is successful. Interactively constructing meaning is a way of *making common* the meaning among interlocutors, which requires that there be something *made* common—the meaning— and something that *makes* common—the co-construction process (Dance 1970). This way of parsing communication is related to the distinction between the *informational* dimension and the *sociopragmatic* dimension, a distinction that has its roots in communication science (e.g., Watzlawick et al. 1967; Fisher 1979). Our development of these dimensions, though, will be slightly different than what is found in the communication literature.

Communication understood in this way needn't involve language, and the co-construction process can leverage other ways of creating the interlocking expectations that are essential to communication (cf. Grice 1957). Nevertheless, communication among collaborators in a cross-disciplinary team will often require language, since without it communicating the complex nuance and detail of collaborative research would be impossible. TDI is primarily focused on linguistic communication, which is central to the design of Toolbox interventions. In what follows, we take the unit of analysis to be a linguistic episode of communication (cf. Keyton et al. 2010).

Consider a linguistic episode L_1 that occurs between times t_1 and t_2 and involves two speakers, S_1 and S_2, who exchange several linguistic utterances, U_1 ... U_n in context C and derive a number of conclusions, P_1 ... P_6, as a result of the interaction. (See Figure 5.2.)

With this toy model, we can begin by distinguishing the *agential* elements of L_1 from its *linguistic* elements. (For communication in general, we would refer to *symbolic* elements rather than just linguistic elements. See Keyton and Beck 2010, p. 335.) The agential elements consist of the agents S_1 and S_2, and also their actions, which include their utterances, thoughts, and their interactions. The last consists of their interlocking utterances and the physical signals and responses they exchange by virtue of physical proximity. The linguistic elements consist of the sentences in their utterances and the contents of their thoughts.

The second distinction we make is between the *content* of the episode and its *form*. The content is the *meaning* contributed over the course of the linguistic episode to the conclusions reached by S_1 and S_2. The form comprises the structures and processes by which the meaning is contributed, which include the language used, the speakers' non-verbal cues, the interactions between the speakers that contribute to their relationship, and contextual aspects that are salient.

The formal study of language provides us with a distinction between content, understood as the meaning of the uttered sentences, and form, which is the grammar

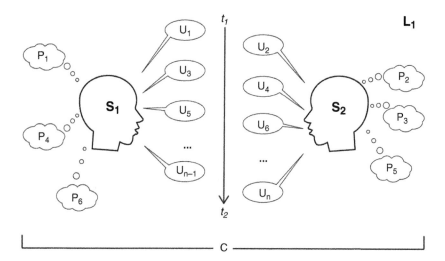

FIGURE 5.2 A formal model of a linguistic episode L_1, comprising two speakers, S_1 and S_2, who exchange a series of utterances U_1 ... U_n in context C and form several conclusions P_1 ... P_6 over a period of time that begins at t_1 and ends at t_2.

(Campbell et al. 2002). Linguistic significance is modeled by semantics (what is encoded in the sentence uttered) and pragmatics (what is contextually implied by the speaker), while grammar is modeled by syntax (Korta and Perry 2020). (See Bianchi 2004 and Szabó 2005 for more of this rich conversation in the philosophy of language and linguistics.) Non-verbal cues are similar to uttered sentences, in that they impart a contextually constrained message (e.g., compare a hand wave when you approach someone with a hand wave when you depart). However, the message can be less conventionally constrained than linguistic messages (Lascarides and Stone 2009). Behaviors intended to establish one's authority and alter the confidence of one's interlocutor are all examples of interactional forms that inflect how one understands the conventionalized content of an episode and also suggest conclusions about one's interlocutor. As for contextual aspects, these are rendered salient by the elements of the episode we have just described, as well as events that occurred prior to L_1 and any prior interactions involving S_1 and S_2 (Keyton and Beck 2010).

Our two distinctions jointly act to bring attention to the two dimensions of communication that matter to us in TDI: the *informational* dimension and the *sociopragmatic* dimension (see Figure 5.3).

The informational dimension is related to what is *made* common and the sociopragmatic dimension to what *makes* common. The informational dimension concerns the content recovered by interlocutors from the language used and from each other and their interactions. This is the content that populates the conclusions P_1 ... P_6 made by S_1 and S_2 in L_1. The sociopragmatic dimension pertains more closely to the formal aspects of the episode, including the actions and interactions of the interlocutors and the formal characteristics of the language used. This dimension accommodates what is typically described as *relational* communication (Keyton 1999), but also includes relationships between speakers and the languages involved. In a typical exchange between speakers using the same language, this is not an important consideration, but it can be important when different languages are used, as we discuss below. Not every aspect of form or content and not every agential

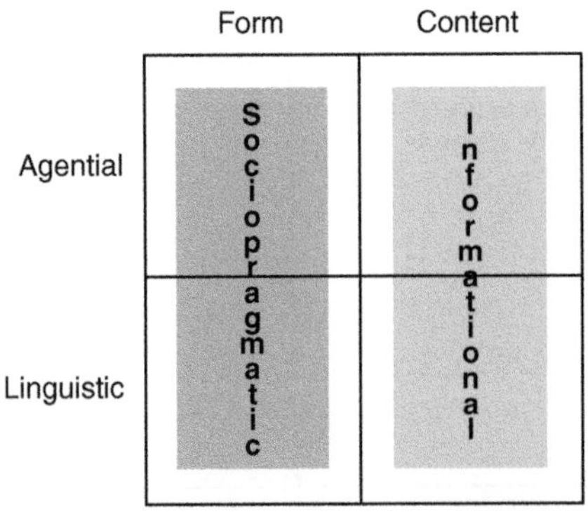

FIGURE 5.3 The sociopragmatic and informational dimensions of communication presented in relation to the aspects of form and content and the agential and linguistic elements in communicative exchanges.

or linguistic element necessarily contributes to the informational or sociopragmatic dimensions of a given communicative episode. Not all will be deemed relevant by the interlocutors.

This dimensional way of thinking about linguistic communication makes it possible to mark the presence of relationships among speakers, but also relationships between a speaker and the language of another speaker, i.e., pragmatic relationships involving both the agential/form quadrant and the linguistic/form quadrant in Figure 5.3. This way of thinking about communication has much in common with other approaches. The informational/sociopragmatic distinction is in line with the content/relationship distinction discussed by Watzlawick et al. (1967), who emphasize that every act of communication has a content aspect and a relational aspect, with the former understood in terms of information and the latter in terms of interpersonal relationships. Fisher (1979) explicitly associates the content dimension with *task* communication and the relational dimensions with *social* communication. (For more on task communication and relational communication see Hirokawa and Salazar 1999 and Keyton 1999.) Our informational dimension reflects the communication perspective of Keyton et al. (2010, p. 275), who note that "team members use information not only to inform but also to persuade and create relationships." Accordingly, the message of a linguistic signal (e.g., one of the U_i's in L_1 shown in Figure 5.2) will rarely be sufficient for the formation of the conclusions on which success of the episode depends. This characteristic of linguistic messages distinguishes this approach from information-based communication theories, such as Shannon (1948). We refer to the dimension as "sociopragmatic" to accommodate both interpersonal and group interrelationships as well as the formal (i.e., structure and process) relationships among speakers, languages, and contexts.

By emphasizing the role that relationships play in communication, our approach is in line with that of Keyton (1999), Keyton and Beck (2010), and Keyton et al. (2010), all who have championed a conception of team communication that emphasizes the interactive, co-construction of meaning. Their work takes the interactive negotiation of meaning to be crucial to communication. Following Schegloff (1991), Keyton and Beck (2010, p. 338) view meaning as being "jointly produced as it resides in the interaction moves of those participating in the conversation." This way of understanding meaning highlights the importance of formal structures and processes involving interacting agents and their utterances. *Relational* communication also figures importantly in their work. Keyton (1999, p. 192) surveys the literature on relational communication, understood as the "affective or expressive dimension of group communication," and presents a model of it that emphasizes relational coordinates at all stages of communication. These include inputs (e.g., relational messages, such as venting emotion), process (e.g., interactions), and outcomes (e.g., cohesiveness). Keyton et al. (2010) supply a communication perspective on team macrocognition, understood to be processes deployed by teams to make knowledge together. This perspective emphasizes the dynamic, negotiated interaction among teammates to make meaning together, unfolding in ways that must be understood both informationally and sociopragmatically.

Craig (1999) provides an overview of several traditions of communication theory that are represented in our approach. The "semiotic" tradition understands communication as the *"intersubjective mediation by signs"* (p. 136). This is easy to recover from L_1, and its emphasis on the importance of a common language is often highlighted in the context of cross-disciplinary work (e.g., Galison 1997). The "phenomenological" tradition understands communication as *"dialogue or experience of otherness"* (p. 138). Communication understood as dialogue is crucial to the workshops we conduct. The "cybernetic" tradition understands communication as *"information processing"* (p. 141), and the "socio-psychological" tradition understands it as *"a process of expression, interaction, and influence"* (p. 143). These traditions align with the two dimensions of communication we have discussed.

Although our approach overlaps each of these traditions, important differences remain. Communication as we understand it need not involve signs; this sets it apart from the semiotic tradition. We take language to be central for communication among cross-disciplinary collaborators, but cross-disciplinary communication can be successful even in the absence of a common technical language among collaborators. According to Craig (1999, p. 138), the phenomenological tradition emphasizes "the interplay of identity and difference in authentic human relationships and cultivates communication practices that enable and sustain authentic relationships," with *dialogue* qualifying as "authentic communication." So understood, dialogue qualifies as a norm for communicative interaction among collaborators, and it is something we try to model in Toolbox workshops (see Chapters 2 and 7). However, communication is a challenge for cross-disciplinary collaborators precisely because so much of it fails to conform to this norm.

Craig notes (p. 142) that the cybernetic tradition emphasizes the transmission of information when the "whole is greater than the sum of the parts." In contrast, the socio-psychological tradition emphasizes the importance of agent relationships to communication—"psychological predispositions," such as emotional states, mediate attempts at communication in ways that are inflected by social processes (p. 143). Both of these traditions emphasize a dimension we take to be central to the process of communication—the informational dimension in the case of cybernetics, and the relational dimension in the case of socio-psychology. However, each tells only part of the story.

5.2.3 COMMUNICATION IN CROSS-DISCIPLINARY CONTEXTS

By communicating with one another, people construct a joint reality that is grounded in the meaning they make together out of their utterances, actions, and interactions. The joint reality is negotiated and dynamic, changing as the contributions change. The model we have proposed for linguistic communication, with informational and sociopragmatic dimensions framing the interaction of agents and languages, helps us identify and understand the meanings that can be made common by being co-constructed, i.e., the content of communication, and the ways in which those things are made common, i.e., the form of communication.

This model is designed to help us understand communication in cross-disciplinary contexts. Communication can be especially challenging in these contexts for a variety of reasons. Some of these challenges are *informational*, as when it is difficult to understand others due to their different training or to their different cultural orientation (Lélé and Norgaard 2005; Roy et al. 2013). Different research worldviews are critical inputs into a cross-disciplinary integration process, but coordinating them takes real work (Ramadier 2004; O'Rourke and Crowley 2013), especially when this requires translating between different technical vocabularies (Bracken and Oughton 2006; Donovan et al. 2015). Other challenges are *sociopragmatic*, such as the difficulty of relating to one's collaborators or their technical languages. Different disciplinary cultures have different epistemic sensitivities and methodological proclivities, which sometimes make it difficult to trust collaborators from different disciplines due to one's own ignorance about their expertise (Campbell 2005; National Research Council 2015; Piso et al. 2016; Hall et al. 2018). The development of the understanding required for trust is made harder by differences in technical vocabularies (although see Choi and Richards 2017).

Being a successful communicator in cross-disciplinary contexts requires *communication competence*. Jablin and Sias (2001, p. 824) discuss this in the group context and focus on the resources a group needs to meet the "information requirements of their information environments." These resources that enable a group to meet its information requirements include (p. 824) "a group's internal feedback structures, intergroup communication networks, and the communication practices and structures that help it maintain successful collaboration."

Thompson (2009) applies this notion to communication in interdisciplinary research teams and adds a group perspective to *communication competence* to obtain *collective communication competence* (CCC). She takes CCC to consist of the abilities of group members (p. 281) "to be simultaneously appropriate and effective"; these abilities are grounded in complex "interrelationships among communicators, contexts, and goals." Thompson asserts that appropriate and effective interdisciplinary communication is aided by spending time together (cf. Hampton and Parker 2011), building trust, talking about the work, and coordinating technical vocabularies. Interrelationships that can be especially influential in helping teams develop CCC are being fully present during communication episodes, being reflexive together, communicating offline about difficulties, and sharing laughter.

Thompson's ethnographic work provides insight into how the sociopragmatic form of communication in a cross-disciplinary setting can help a team meet its information requirements, but not all of the items she emphasizes are specifically related to the circumstances of *cross-disciplinary* contexts. We add a fourth "C"—*cross-disciplinary*—so that we can focus on the conditions that are conducive to *cross-disciplinary collective communication competence* (4C). We organize these conditions into informational and sociopragmatic groups, and we develop the conditions as recommendations about communication processes to utilize in cross-disciplinary contexts.

The 4C informational conditions concern co-constructed meaning that influences the conclusions interlocutors reach in conversational episodes like L_1 above. As Thompson notes, differences among cross-disciplinary research and practice worldviews can be obstacles to meaningful discourse. The cross-disciplinary literature highlights several conditions that can help collaborators with different perspectives *localize* those perspectives for one another (cf. Crowley et al. 2010).

We emphasize three of these conditions that are conducive to cross-disciplinary collective communication competence:

1. *Attend to differences in technical language.* Collaborators should cultivate an understanding of the different technical languages employed in a project and develop an adequate understanding of relevant technical terms and ways of speaking (Wear 1999; Klein 2012). What counts as an "adequate understanding" will vary, depending on the history of the group and its context—sometimes it is necessary to work on developing this understanding from the start (e.g., von Wehrden et al. 2019), although sometimes it is not needed at all (Choi and Richards 2017).
2. *Cultivate critical reflexivity.* Critical reflexivity in a team is the capacity to reflect on team knowledge, commitments, values, and priorities with a view to improving team function (Salazar et al. 2019). As Thompson (2009, p. 287) notes, reflexive communication requires "a subjective process of self-consciousness inquiry and an awareness of social behavior and its impact on group dynamics," which can enhance team effectiveness by revealing problematic aspects of team performance.
3. *Encourage perspective seeking.* Perspective-seeking behavior is the effort to adopt the perspective of another and see the world through their eyes (Hoever et al. 2012; Looney et al. 2014). When teammates think about their common problem from each other's perspective, it can reveal differences in emphasis, coverage, or priority, and engender empathy that can help make differences in meaning more apparent.

The 4C *sociopragmatic conditions* address poise and practice that can help create a cooperative culture. Core beliefs and values differ in cross-disciplinary contexts, and so collaboration can be stressful, especially when career prospects and professional reputation are involved. Creating a communication dynamic that supports coordination and cooperation, as opposed to competition, can help teammates find solutions to the problems that arise. Three conditions that encourage cooperative interaction are worth emphasizing:

1. *Encourage intellectual humility.* Cross-disciplinary research brings together people with different expertise, which means that no one person has all of the required expertise and knowledge. A crucial attitude for collaborators is intellectual humility (Alfano et al. 2017; Wright et al. 2017), which supports critical deference to one's colleagues about topics of which one is not an expert and also an openness to learning new perspectives (Kjellberg et al. 2018).
2. *Concentrate on building trust.* Trust is a key determinant of success in any collaborative context, but it is especially important in cross-disciplinary contexts (Maglaughlin and Sonnenwald 2005; cf. Whyte and Crease 2010). Sustainable trust in a team should be interpersonal, but this requires opportunities to do enough meaningful work together to cultivate the *interactional* expertise necessary to support conversations about integrative possibilities (Collins and Evans 2002).

3. *Acknowledge joint construal.* Meanings that emerge through cross-disciplinary communication are typically co-constructed (Hall and O'Rourke 2014). As such, project conclusions are products of a process of *joint construal* (Clark 1996, p. 212) in which team members "settle on what speakers mean." Acknowledging the cooperative creation of project meaning encourages "we" thinking among team members that can help them surmount difficulties (cf. Tuomela 2007).

5.3 INTEGRATION

Integration is crucial for successful cross-disciplinary research and practice, especially in a collaborative environment (Klein and Newell 1996; National Academy of Sciences 2005; Pohl et al. 2008; Ratcheva 2009; Bergmann et al. 2012; Klein 2012; Repko 2012; Bammer 2013; Gerson 2013). Bergmann et al. (2012, p. 42) remark that "the importance of integration work can hardly be overestimated for transdisciplinary research," and Newell (2001, p. 2) notes, "By definition, interdisciplinary study draws insights from relevant disciplines and integrates those insights into a more comprehensive understanding." Cooperation is the foundation of cross-disciplinary integration. In a collaborative, cross-disciplinary context, integration requires cooperation because it is complicated, involving purposely bringing diverse inputs from different collaborators together into the complex responses required to address complex problems. Klein (2012, p. 284) underscores the connection between integration and cooperation, noting that "Integrative capacity and the ability to work in teams are coupled increasingly across all sectors of our lives." In other words, where teams must engage in integrative research or practice, integrative success will often be aided by achieving the kind of cooperation required for well-oiled teams.

Integration is also hard (Lynch 2006; Newell 2007). The challenges of integration are both conceptual and practical. It is often unclear precisely what integration *is*—the *conceptual* challenge—and *how* to go about it—the *practical* challenge. In what follows next, we address the conceptual challenge and thereby support cross-disciplinary research practitioners looking to address the practical challenge. Our goals are (a) to spell out what cross-disciplinary integration is in more detail by drawing on literatures in interdisciplinarity and philosophy of science, and (b) to indicate how integration, so understood, can be challenging in cross-disciplinary contexts.

5.3.1 What Is Integration?

We understand integration in accordance with the Input-Process-Output (IPO) heuristic (O'Rourke et al. 2016). Integration is a "bringing together" of things to make something new and different. In combining inputs, integration often produces an output that is more than the sum of the inputs; the output is something new. The inputs may fit together easily or be in conflict. The process can be dynamic with various phases (Hall et al. 2012), iterative (Newell 2007; Bergmann et al. 2012; Hall et al. 2012; Klein 2012; Repko 2012), or recursive (Bergmann et al. 2012). The output can be reductive or non-reductive (Grantham 2004). Input, process, and output can be conceived of descriptively (i.e., what each *is*) or normatively (i.e., what each *should* be).

In addition, integration in cross-disciplinary research is purposeful (Brigandt 2010; Bammer 2013). It doesn't happen by accident, and haphazard combination of inputs is insufficiently purposeful to be integrative. (Of course, though, there can be unplanned combinations of insights that contribute to it.) Some particular research question, problem, or purpose spurs the process, guiding the selection and combination of inputs and grounding evaluation of the output(s) of the integration.

It is helpful to develop a taxonomy of integration based on the IPO model of O'Rourke et al. (2016). Our epistemic taxonomy classifies kinds of integration based on what is known and what is unknown about the inputs, process(es), and output(s) when integration starts. If one knows what kind of integration to pursue, this taxonomy can be used to better anticipate the challenges that will arise. For example, suppose that a cross-disciplinary team is initially unaware of what inputs they will need to integrate, how they will go about integrating them, and what the integrated output should do. Such a team faces different challenges than one that is only ignorant of the content of the integrated output it seeks to produce. These two teams would attend to different aspects of the taxonomy as they work through their cross-disciplinary projects.

To illustrate the taxonomy, consider an analogy with the use of LEGO® building bricks, which can be assembled in a variety of ways. For example, one can buy a pre-designed set that comes with instructions and only the pieces needed to build what is pictured on the box. The input, process, and output are clearly defined. Alternatively, one can build something without a clear output in mind, a defined process, or a set of predetermined pieces. In between knowing everything in advance and knowing nothing in advance lies a wide range of other ways to assemble LEGO bricks. If there are two possible states (known or unknown) for each of the three components to integration (input, process, and output), then there are eight possible kinds of integration for cross-disciplinary research (and LEGO building!), as shown in Table 5.1. Fam and Sofoulis (2017) contrast the multitude of integrative complexities they faced in an cross-disciplinary project with "a positivist 'knowledge integration' perspective" where "different knowledges are like [LEGO] blocks that can be readily combined with other 'data sets' or 'skill sets' to create a unified structure." As we discuss and Table 5.1 demonstrates, we have something considerably more complex and non-positivistic in mind in this taxonomy of cross-disciplinary and LEGO integration types.

Different kinds of integration present different challenges based on what is unknown. To continue the LEGO analogy, some ways of building begin with known inputs; the pieces to be used are already determined (alpha – delta). When the inputs are unknown (epsilon – theta), there are many possible blocks that could be used, and you don't know at the outset which will end up in the final product. This complicates the building and nearly ensures that many pieces will be tried, removed, and perhaps replaced. The process takes longer.

When the process(es) are unknown, the means for achieving integration is not determined at the outset. Under these circumstances, integration typically requires that the process(es) be iterative, especially if the output is also unknown. You might

TABLE 5.1
Taxonomy of Integration Based on What Inputs, Processes, and Outputs Are Known at the Outset

Name	Inputs	Processes	Outputs	Description
alpha	Known	Known	Known	*Pre-packed set*: All parts, instructions, and final creation predetermined
beta	Known	Known	Unknown	*Mystery build*: Take the provided parts and follow the instructions to create a surprise
gamma	Known	Unknown	Known	*Figure-out-how-to-build-it challenge*: The output can be created from the provided inputs; the challenge is to figure out how; possibility for multiple process to produce the specified output
delta	Known	Unknown	Unknown	*What-can-you-make-with-these challenge*: Specific inputs are provided as a creativity challenge; many processes and outputs are possible
epsilon	Unknown	Known	Known	*What is this made from?* Inputs unspecified; process defined vaguely enough to work for a variety of inputs; output is generic and not specific to particular inputs
zeta	Unknown	Known	Unknown	*Process only*: Process defined vaguely enough to work for a variety of inputs and can create a variety of outputs depending on the inputs
eta	Unknown	Unknown	Known	*See-it-design-yourself challenge*: One must figure out how to create the final output and what parts to use
theta	Unknown	Unknown	Unknown	*Free design*: Everything unspecified

try one LEGO building technique (i.e., one process), realize it won't produce the desired result, and so go back and try a new technique. For LEGO, when the process is known (e.g., because instructions are provided), inputs are typically specified and so integration types epsilon and zeta (unknown inputs, known process) are rare. In contrast, for integration in cross-disciplinary research, types epsilon and zeta are common. A plethora of processes exist that can contribute to cross-disciplinary integration, including many social processes, for instance, that purport to turn nearly any set of individuals into a team, including the method Holton (2001) describes for building teams in virtual environments and the many ways to foster team integration found in the field of sports psychology (e.g., Fransen et al. 2015).

Last, in the taxonomy provided in Table 5.1, the output of integration is to be understood normatively as establishing the success criteria for the integrative episode.

If the output is known at a fine level of detail, then integrative success will depend on generating that product in particular. When building from a pre-packaged LEGO set, what is on shown on the box defines the criterion for success. When the output is not so precisely known, it is harder to define success. Of course, because integration is a purposeful process, it is known in each case that there is a goal, which ensures that there are success criteria. The existence of success criteria entails once again that not just any integrated result would count as successful, but knowing that there are success criteria is compatible with not knowing what they are.

The primary point of the taxonomy in Table 5.1 is to emphasize that what you know and don't know when you begin can be used to classify integration into different types. When engaging in cross-disciplinary research, alpha integration is different than theta integration. Accordingly, cross-disciplinary research teams should identify which sort of integration they are embarking upon. Doing so will allow them to recognize the kinds of difficulty they face, marshal the appropriate set of resources, and set appropriate expectations.

Two significant oversimplifications must be noted. First, the taxonomy depicts inputs, process, and outputs as completely known or completely unknown. Knowledge, however, comes in degrees: you can know all of something or none of it, but often you are somewhere in between. Some but not all of the inputs might be known, some but not all of the process might be known, and some but not all of the aspects of the final output might be known. It is more precise to regard "known" and "unknown" as indicating whether each component is mostly known or mostly unknown. For example, alpha integration occurs when you know most of the inputs, you are sure about most of the steps in the process, and you are sure of most details of the output. Conversely, when only one of many inputs is known, only the first stage of the process is spelled out, and the output is murky at best, that is theta integration.

The second oversimplification stems from drawing a parallel between integration in cross-disciplinary research and LEGO integration. LEGO bricks are designed to fit together, even when purchased decades apart. The inputs in integration in cross-disciplinary research, by contrast, are not designed to fit together and might not be able to be integrated at all. In this context, as Newell (2007, p. 256) tells us:

> Integration would consist, as is too often supposed, of putting together a jigsaw puzzle. Instead, integration is more like discovering that many of the jigsaw pieces overlap, and worse, that the pieces seem to come from different jigsaw puzzles and that many of the pieces are missing. The decisions one makes about how to modify the pieces determine if and how the puzzle can be sufficiently solved, in spite of the missing pieces, to make out the picture.

In this respect, cross-disciplinary research integration is much more like taking disparate objects and creating a new way for them to fit together that was never designed or imagined. The LEGO comparison is nevertheless useful because it makes it easier to understand each of the kinds of integration presented in Table 5.1. Once the taxonomy is understood, recognizing the eight kinds of cross-disciplinary research integration is easier. By identifying the kind of integration one is pursuing, one can better foresee the challenges ahead.

5.3.2 SOCIAL AND EPISTEMIC INTEGRATION

We can also classify types of cross-disciplinary research integration in terms of the distinction between social and epistemic integration (O'Rourke et al. 2016, 2019). Bergmann et al. (2012, p. 45) describe "social" integration as "a matter of differentiating and correlating the participating researchers' different interests and activities, as well as of the sub-projects or organizational units." Social integration is the process of transforming a collection of separate voices saying, "I…" into a chorus that says, "We…." The inputs are people and the output is a team that is more than the sum of its parts. Who will make up the team is sometimes clear from the outset, but not always, and the team-building process can be known initially, but not always. For example, if a group decides at the outset to contract with TDI to run workshops, they know that these workshops will be a part of the process that transforms them into a team. For social integration, the output will be a team, but the specific contours of that team and the roles played on it may be unknown when the integrative process begins.

If a socially integrated team is a chorus, epistemic integration provides the lyrics and score. Bergmann et al. (2012, p. 45) describe epistemic integration as "a matter of understanding the methods and terms of other disciplines; clarifying the limits of one's own knowledge; and developing methods and building theories together." Epistemic integration takes a diverse set of epistemic inputs and produces integrated epistemic outputs. These integrated epistemic outputs are either what the team now knows and can communicate outside the team or the mutual understanding that the team developed in order to generate this sharable knowledge. The epistemic inputs and outputs can be concepts (Eigenbrode et al. 2007), data (Leonelli 2013), explanations (Brigandt 2013), knowledge (Zierhofer and Burger 2007), methods (Tress et al. 2004), norms and values (Eigenbrode et al. 2007; Cosens et al. 2011), or goals. Not each of these items must be included in epistemic integration, but all of them are potentially relevant. The production of integrated output is aided by mutual understanding of the inputs on the part of all of the researchers, which is attained by the act of "seeing through the eyes of others" (Looney et al. 2014). Mutual understanding enables teams to communicate more effectively and to discover differences and overlap, make modifications, and develop ways to fit their epistemic inputs into an integrated output. Such an output is collectively held by the team and not necessarily by any single team member. This is apparent for teams that are very large (e.g., Aad et al. 2015), in which there may be no individual who understands the entirety of an article the team has written. In such a case, the *team* is the entity that possesses the co-constructed knowledge.

Though conceptually separate, these two kinds of integration are closely related in practice in two ways. First, since the integrated output in epistemic integration is properly attributable to the team as a whole (and not necessarily to any of the individuals on the team), there must first be a team to hold that integrated worldview. So, epistemic integration in cross-disciplinary research typically requires social integration. (On occasion, a savvy individual may "lead the way" to a unified worldview even though the team is divided; however, the identification by the team that the integrated worldview is "ours" requires at least some social integration.) Second, epistemic integration can foster social integration. A group of individuals figuring out how to relate and integrate their disparate research worldviews is a means by which the "we" of a team can be constructed.

5.3.3 CHALLENGES

A cross-disciplinary team may not understand or agree on what type and amount of integration is needed in order for their project to succeed (Klein 2012). This lack of clarity is often the biggest challenge to successful integration in cross-disciplinary research (Leonelli 2013). Our taxonomy of integration can assist in addressing this challenge. When initiating the process of integration, it helps to determine how much is known about the inputs, process(es), and output(s). By doing so, any gaps of knowledge about the inputs, process(es), and output(s) become clear. Eliminating these gaps can then serve as a project starting point, although it is possible that some of these gaps will never be eliminated. For instance, the best epistemic outputs for a given episode of integration may remain unknown. Nevertheless, learning as much as is practical about the inputs, processes, and outputs is important to overcoming the biggest challenges to integration in cross-disciplinary research.

Resolving some of the unknowns can be particularly daunting. In epistemic integration, the inputs can be difficult to identify. Since it can be unclear which potential epistemic contributions of each individual team member will be relevant, epistemic integration is often one of the epsilon through theta varieties. It may be abstractly apparent which skills or perspectives are needed from each team member—e.g., the biostatistician is needed to design the experiment, the bioengineer to design the implant, and the transplant surgeon to install and monitor it. But the specific contributions of these collaborators may not be known ahead of time, or even well into the development of the project. As the project unfolds, the specific contributions will become clearer and the project will become one of the alpha through delta varieties.

Furthermore, an individual's set of epistemic items is typically vast and complex. When teams begin to create an integrated epistemic output, team members may themselves initially be unaware of some of their own relevant epistemic inputs, such as their philosophical assumptions. One's worldview often includes unrecognized philosophical assumptions on a wide range of topics, concepts, and practices (O'Rourke and Crowley 2013). For example, the conceptual basis of hypotheses often goes unanalyzed as scientists go about their quotidian work. If a team member has never examined and understood their views on the nature and function of hypotheses in scientific research, then gaining this self-understanding is a necessary precursor to the integrative task of understanding the scientific worldviews on a team and how they all differ, which is itself necessary for the team to produce an integrated epistemic output (Donovan et al. 2015).

The process of creating a team from disparate individuals needn't be especially mysterious. Methods to do so have been developed in a wide array of contexts, including sports, the military, and business. In addition, several methods have been developed specifically for cross-disciplinary research (cf. Eigenbrode et al. 2007;

TABLE 5.2
The Different Inputs, Processes, and Outputs for Social Integration and Epistemic Integration

	Inputs	Process(es)	Output(s)
Social integration	Individuals	TDI workshops Team-building exercises (e.g., trust falls) Goal articulation and reinforcement Develop communication strategies or technologies	Team
Epistemic integration	Methods, language/ terminology, concepts, norms, values, explanations, data, knowledge, or goals	Internal communication TDI workshops Co-authoring Any process that fosters mutual understanding	Integrated methods, language/ terminology, concepts, norms, values, explanations, data, knowledge, or goals

National Research Council 2015; Wargad and Puri 2015; Bennett et al. 2018). Nonetheless, the process of epistemic integration can be quite vexing. The challenge is to bridge potentially wide gaps in epistemologies and/or ontologies among team members. When engaged in cross-disciplinary research, the epistemic inputs are different for every team, so there can be no one road map (Klein 2012). Repko (2012) highlights the difficulty of this challenge as follows (p. 263): "Integration is something that we must create. It does not derive from a pre-determined pattern or universal template that is applicable beyond the specific problem, issue, or question being addressed in a particular course or research project." See Table 5.2. There is no simple, easy way to integrate the diverse epistemic inputs of a cross-disciplinary research team. How to best integrate will be determined by the specifics of each team and each project. Social factors—such as power dynamics within a team or external social pressures—only complicate these matters further.

5.4   THE INTERACTION BETWEEN COMMUNICATION AND INTEGRATION

In order to be successful, a cross-disciplinary research team must acquire "cross-disciplinary collective communication competence" (4C). Doing so is no easy task. It requires an awareness of the diverse audiences and modes of cross-disciplinary research communication. 4C also requires awareness of where integration is needed and a facility at pursuing integrative objectives. The different kinds of cross-disciplinary research communication and the different modes of integration shown above indicate the complexity of 4C, thereby providing cross-disciplinary research teams with a better appreciation of what is required to attain it.

In this chapter, we have distinguished between informational and sociopragmatic communication. It is relatively straightforward to see a correspondence between these two kinds of communication and epistemic and social integration, with informational communication being crucial to epistemic integration and sociopragmatic communication facilitating social integration. The depth of the relationship between integration and communication can be plumbed more deeply, however. Understanding integration according to the Input-Process-Output heuristic, we can ask whether and how communication contributes to the inputs to be integrated, the integrative process, or the integrated output. Communication will typically support integration, but in different ways. Consider the differences in integration between communication *in*, *out*, and *across*. As a cross-disciplinary research team works to achieve epistemic integration, communication *in* serves as part of the process. The more a team works to become aware of its internal communication dynamic, the more communication *in* will contribute to alpha, beta, epsilon, and zeta types of integration.

Integration involving communication *across* and communication *out* typically takes different forms. Once a cross-disciplinary research team produces an integrated output, they may want to communicate that output *across* disciplinary lines to other academics or *out* to non-academics for whom the research may be relevant. Cross-disciplinary research teams often intend their integrated outputs to become inputs for epistemic integration by others. Accordingly, the integrated outputs they communicate across or out may initiate further integration. The cross-disciplinary research team's output (e.g., presentations, scholarly articles, reports, videos, interviews) becomes an input in the new integrative process, along with relevant aspects of the intended audience's philosophical worldview, methods, knowledge, and so forth. The cross-disciplinary research team selects their intended audience, informs that audience of the team's output, and invites the audience to take this output and integrate it with their own inputs. Since the inputs are largely defined ahead time, communication across or out typically involves alpha, beta, gamma, or delta types of integration. The process begins with communication by the cross-disciplinary research team and may involve considerable back-and-forth between the team and their audience. As we have noted, the means of communicating across will often differ from those of communicating out.

Focusing specifically on communication *in*, the kind of integration a team is engaged in or embarking upon will strongly influence the nature of their internal communication—or at least how they *should* communicate. The more that a team does not known about the inputs, process, or output, the more questions team members will need to ask one another. For example, when a team does not know at the outset how to integrate (integration types gamma, delta, eta, or theta), team members will ask questions like, "What should we do next?" Considerable communication *in* will be needed to answer this question. If, however, the process is known (integration types alpha, beta, epsilon, or zeta), communication *in* will typically be more about explaining "Here's what's next." When the inputs for epistemic integration are unknown (integration types epsilon, zeta, eta, or theta), teams will spend time trying to learn about and from one another. For example, they might ask each other about their methods and philosophical views on replication or hypotheses. Finally, when the outputs are unknown (integration types beta, delta, zeta, or theta), team members will need to ask each other whether the output solves the problem or answers the questions that prompted the integration in the first place. Sometimes this remains unclear even at the end of an integrative process. When the output is known, these questions and conversations do not arise.

We close with recommendations that combine our interest in communication and integration. First, consider your audience, as this will define whether you're engaged in communication *in*, *across*, or *out* and will determine what sort of integrative potential your communication may have. Second, consider *who* is doing the integrating. The *who* may be the team members as they engage in social integration or epistemic integration. On the other hand, once a team communicates its output across or out, the *who* could include others who take the team's output and integrate it with their own worldviews, methods, and goals. Last, consider how many of the inputs, processes, and outputs are known to the team members at the outset, as this will influence the nature of the communication in which a team engages.

5.5 ACKNOWLEDGMENTS

We would like to thank Graham Hubbs and Steven Orzack for their editing assistance. O'Rourke's work on this chapter was supported by the USDA National Institute of Food and Agriculture, Hatch Project 1016959.

5.6 LITERATURE CITED

Aad, G., B. Abbott, J. Abdallah, O. Abdinov, R. Aben, M. Abolins, O. S. AbouZeid, H. Abramowicz, H. Abreu, R. Abreu, et al. 2015: Combined measurement of the Higgs boson mass in pp collisions at $\sqrt{s} = 7$ and 8 TeV with the ATLAS and CMS experiments. Physical Review Letters 114:191803.

Alfano, M., K. Iurino, P. Stey, B. Robinson, M. Christen, F. Yu, and D. Lapsley. 2017: Development and validation of a multi-dimensional measure of intellectual humility. PLOS ONE 12:e0182950.

Bammer, G. 2013: Disciplining Interdisciplinarity: Integration and Implementation Sciences for Researching Complex Real-World Problems. Australian National University Press, Canberra.

Bennett, L. M., H. Gadlin, and C. Marchard. 2018: Collaborative Team Science: A Field Guide. National Cancer Institute, Bethesda, MD.

Bergmann, M., T. Jahn, T. Knobloch, W. Krohn, C. Pohl, and E. Schramm. 2012: Methods for Transdisciplinary Research. Campus Verlag, Frankfurt.

Bianchi, C., ed. 2004: The Semantics/Pragmatics Distinction. CSLI Publications, Stanford, CA.

Bosque-Pérez, N. A., P. Z. Klos, J. E. Force, L. P. Waits, K. Cleary, P. Rhoades, S. M. Galbraith, A. L. B. Brymer, M. O'Rourke, S. D. Eigenbrode, B. Finegan, J. D. Wulfhorst, N. Sibelet, and J. D. Holbrook. 2016: A pedagogical model for team-based, problem-focused interdisciplinary doctoral education. BioScience 66:477–488.

Bracken, L. J., and E. A. Oughton. 2006: "What do you mean?" The importance of language in developing interdisciplinary research. Transactions of the Institute of British Geographers 31:371–382.

Brigandt, I. 2010: Beyond reduction and pluralism: Toward an epistemology of explanatory integration in biology. Erkenntnis 73:294–311.

———. 2013: Integration in biology: Philosophical perspectives on the dynamics of interdisciplinarity. Studies in History and Philosophy of Biological and Biomedical Sciences 44:461–465.

Brod, S. 2014: *The importance of science communication*. Nature Jobs. Downloaded from http://blogs.nature.com/naturejobs/2014/09/04/the-importance-of-science-communication/.

Campbell, J. K., M. O'Rourke, and D. Shier. 2002: Investigations in philosophical semantics. Pp. 1–17 *in* J. K. Campbell, M. O'Rourke, and D. Shier, eds. Meaning and Truth: Investigations in Philosophical Semantics. Seven Bridges Press, New York.

Campbell, L. M. 2005: Overcoming obstacles to interdisciplinary research. Conservation Biology 19:574–577.

Choi, S., and K. Richards. 2017: Interdisciplinary Discourse. Palgrave Macmillan, London.

Clark, H. H. 1996: Using Language. Cambridge University Press, Cambridge.

Collins, H. M., and R. Evans. 2002: The third wave of science studies: Studies of expertise and experience. Social Studies of Science 32:235–296.

Cosens, B., F. Fiedler, J. Boll, L. Higgins, B. K. Johnson, E. Strand, P. Wilson, M. Laflin, R. Szostak, and A. Repko. 2011: Interdisciplinary methods in water resources. Issues in Interdisciplinary Studies 29:118–143.

Craig, R. T. 1999: Communication theory as a field. Communication Theory 9:119–161.

Crowley, S. J., S. D. Eigenbrode, M. O'Rourke, and J. D. Wulfhorst. 2010: Cross-disciplinary localization: A philosophical approach. MultiLingual 114:1–4.

Dance, F. E. X. 1970: The "concept" of communication. Journal of Communication 20:201–210.

Dietz, T. 2013: Bringing values and deliberation to science communication. Proceedings of the National Academy of Sciences 110:14081–14087.

Donovan, S. M., M. O'Rourke, and C. Looney. 2015: Your hypothesis or mine? Terminological and conceptual variation across disciplines. SAGE Open 5:1–13.

Eigenbrode, S. D., M. O'Rourke, J. D. Wulfhorst, D. M. Althoff, C. S. Goldberg, K. Merrill, W. Morse, M. Nielsen-Pincus, J. Stephens, L. Winowiecki, and N. A. Bosque-Pérez. 2007: Employing philosophical dialogue in collaborative science. BioScience 57:55–64.

Falk-Krzesinski, H. J., N. Contractor, S. M. Fiore, K. L. Hall, C. Kane, J. Keyton, J. T. Klein, B. Spring, D. Stokols, and W. Trochim. 2011: Mapping a research agenda for the science of team science. Research Evaluation 20:145–158.

Fam, D., and Z. Sofoulis. 2017: A 'knowledge ecologies' analysis of co-designing water and sanitation services in Alaska. Science and Engineering Ethics 23:1059–1083.

Fiore, S. M., M. A. Rosen, K. A. Smith-Jentsch, E. Salas, M. Letsky, and N. Warner. 2010: Toward an understanding of macrocognition in teams: predicting processes in complex collaborative contexts. Human Factors 52:203–224.

Fisher, B. A. 1979: Content and relationship dimensions of communication in decision-making groups. Communication Quarterly 27:3–11.

Fransen, K., S. A. Haslam, N. K. Steffens, N. Vanbeselaere, B. De Cuyper, and F. Boen. 2015: Believing in "us": Exploring leaders' capacity to enhance team confidence and performance by building a sense of shared social identity. Journal of Experimental Psychology: Applied 21:89–100.

Galison, P. 1997: Image and Logic: A Material Culture of Microphysics. University of Chicago Press, Chicago.

Gerson, E. M. 2013: Integration of specialties: An institutional and organizational view. Studies in History and Philosophy of Science Part C: Studies in History and Philosophy of Biological and Biomedical Sciences 44:515–524.

Grantham, T. A. 2004: Conceptualizing the (dis)unity of science. Philosophy of Science 71:133–155.

Grice, H. P. 1957: Meaning. The Philosophical Review 66:377–388.

———. 1989: Studies in the Way of Words. Harvard University Press, Cambridge, MA.

Hall, K. L., A. L. Vogel, B. A. Stipelman, D. Stokols, G. Morgan, and S. Gehlert. 2012: A four-phase model of transdisciplinary team-based research: goals, team processes, and strategies. Translational Behavioral Medicine 2:415–430.

Hall, T. E., and M. O'Rourke. 2014: Responding to communication challenges in transdisciplinary sustainability science. Pp. 119–139 *in* K. Huutoniemi and P. Tapio, eds. Heuristics for Transdisciplinary Sustainability Studies: Solution-Oriented Approaches to Complex Problems. Routledge, Oxford.

Hall, T. E., Z. Piso, J. Engebretson, and M. O'Rourke. 2018: Evaluating a dialogue-based approach to teaching about values and policy in graduate transdisciplinary environmental science programs. PLOS ONE 13:e0202948.

Hampton, S. E., and J. N. Parker. 2011: Collaboration and productivity in scientific synthesis. BioScience 61:900–910.

Hirokawa, R. Y., and A. J. Salazar. 1999: Task-group communication and decision-making performance. Pp. 167–191 *in* L. R. Frey, D. Gouran, and M. S. Poole, eds. The Handbook of Group Communication Theory and Research. SAGE, Thousand Oaks, CA.

Hoever, I. J., D. Van Knippenberg, W. P. Van Ginkel, and H. G. Barkema. 2012: Fostering team creativity: Perspective taking as key to unlocking diversity's potential. Journal of Applied Psychology 97:982–996.

Holton, J. A. 2001: Building trust and collaboration in a virtual team. Team Performance Management 7:36–47.

Jablin, F. M., and P. M. Sias. 2001: Communication competence. Pp. 819–864 *in* F. M. Jablin and L. L. Putname, eds. The New Handbook of Organizational Communication: Advances in Theory, Research, and Methods. SAGE, Thousand Oaks, CA.

Keyton, J. 1999: Relational communication in groups. Pp. 192–222 *in* L. R. Frey, D. Gouran, and M. S. Poole, eds. The Handbook of Group Communication Theory and Research. SAGE, Thousand Oaks, CA.

Keyton, J., and S. J. Beck. 2010: Perspectives: Examining communication as macrocognition in STS. Human Factors 52:335–339.

Keyton, J., D. J. Ford, and F. I. Smith. 2008: A mesolevel communicative model of collaboration. Communication Theory 18:376–406.

Keyton, J., S. J. Beck, and M. B. Asbury. 2010: Macrocognition: a communication perspective. Theoretical Issues in Ergonomics Science 11:272–286.

Kjellberg, P., M. O'Rourke, and D. O'Connor-Gomez. 2018: Interdisciplinarity and the undisciplined student: Lessons from the Whittier Scholars Program. Issues in Interdisciplinary Studies 36:34–65.

Klein, J. T. 2012: Research integration: A comparative knowledge base. Pp. 283–298 *in* A. F. Repko, W. H. Newell, and R. Szostak, eds. Interdisciplinary Research: Case Studies of Integrative Understandings of Complex Problems. SAGE, Thousand Oaks, CA.

Klein, J. T., and W. H. Newell. 1996: Advancing interdisciplinary studies. Pp. 393–415 *in* Handbook of the Undergraduate Curriculum: Comprehensive Guide to Purposes, Structures, Practices, and Change. Jossey-Bass, San Francisco.

Korta, K., and J. Perry. 2020: *Pragmatics*. Stanford Encyclopedia of Philosophy. Downloaded from https://plato.stanford.edu/archives/spr2020/entries/pragmatics/.

Kozlowski, S. W. J. 2015: Advancing research on team process dynamics: Theoretical, methodological, and measurement considerations. Organizational Psychology Review 5:270–299.

Lascarides, A., and M. Stone. 2009: A formal semantic analysis of gesture. Journal of Semantics 26:393–449.

Lélé, S., and R. B. Norgaard. 2005: Practicing interdisciplinarity. BioScience 55:967–975.

Leonelli, S. 2013: Integrating data to acquire new knowledge: Three modes of integration in plant science. Studies in History and Philosophy of Science Part C: Studies in History and Philosophy of Biological and Biomedical Sciences 44:503–514.

Looney, C., S. Donovan, M. O'Rourke, S. J. Crowley, S. D. Eigenbrode, L. Rotschy, N. A. Bosque-Pérez, and J. D. Wulfhorst. 2014: Seeing through the eyes of collaborators: Using Toolbox workshops to enhance cross-disciplinary communication. Pp. 220–243 *in* M. O'Rourke, S. J. Crowley, S. D. Eigenbrode, and J. D. Wulfhorst, eds. Enhancing Communication and Collaboration in Interdisciplinary Research. SAGE, Thousand Oaks, CA.

Lynch, J. 2006: It's not easy being interdisciplinary. International Journal of Epidemiology 35:1119–1122.

Maglaughlin, K. L., and D. H. Sonnenwald. 2005: Factors that impact interdisciplinary scientific research collaboration: Focus on the natural sciences in academia. Proceedings of International Society for Scientometrics and Informatrics (ISSI) 2005 Conference:1–12.

McCright, A. M., and R. E. Dunlap. 2011: Cool dudes: The denial of climate change among conservative white males in the United States. Global Environmental Change 21:1163–1172.

McCright, A. M., K. Dentzman, M. Charters, and T. Dietz. 2013: The influence of political ideology on trust in science. Environmental Research Letters 8:44029.

National Academy of Sciences. 2005: Facilitating Interdisciplinary Research. National Academies Press (US), Washington DC.

National Research Council. 2015: Enhancing the Effectiveness of Team Science. N. J. Cooke and M. L. Hilton, eds. National Academies Press, Washington DC.

Newell, W. H. 2001: A theory of interdisciplinary studies. Issues in Interdisciplinary Studies 19:19–25.

———. 2007: Decision-making in interdisciplinary studies. Pp. 245–264 *in* G. Morçöl, ed. Handbook of Decision Making. CRC/Taylor and Francis, Boca Raton, FL.

O'Rourke, M., and S. J. Crowley. 2013: Philosophical intervention and cross-disciplinary science: The story of the Toolbox Project. Synthese 190:1937–1954.

O'Rourke, M., S. J. Crowley, and C. Gonnerman. 2016: On the nature of cross-disciplinary integration: A philosophical framework. Studies in History and Philosophy of Science Part C: Studies in History and Philosophy of Biological and Biomedical Sciences 56:62–70.

O'Rourke, M., S. J. Crowley, B. K. Laursen, B. Robinson, and S. E. Vasko. 2019: Disciplinary diversity in teams, integrative approaches from unidisciplinarity to transdisciplinarity. Pp. 21–46 *in* K. A. Kall, A. L. Vogel, and R. T. Croyle, eds. Advancing Social and Behavioral Health Research through Cross-Disciplinary Team Science: Principles for Success. Springer, Berlin.

Piso, Z., M. O'Rourke, and K. C. Weathers. 2016: Out of the fog: Catalyzing integrative capacity in interdisciplinary research. Studies in History and Philosophy of Science Part A 56:84–94.

Pohl, C., L. van Kerkhoff, G. Hirsch Hadorn, and G. Bammer. 2008: Integration. Pp. 411–424 *in* G. H. Hadorn, H. Hoffman-Riem, S. Biber-Klemm, D. J. W. Grossenbacher-Mansuy, C. Pohl, U. Wiesmann, and E. Zemp, eds. Handbook of Transdisciplinary Research. Springer, Berlin.

Ramadier, T. 2004: Transdisciplinarity and its challenges: the case of urban studies. Futures 36:423–439.

Ratcheva, V. 2009: Integrating diverse knowledge through boundary spanning processes–The case of multidisciplinary project teams. International Journal of Project Management 27:206–215.

Repko, A. F. 2012: Interdisciplinary Research: Process and Theory. SAGE, Thousand Oaks, CA.

Roy, E. D., A. T. Morzillo, F. Seijo, S. M. W. Reddy, J. M. Rhemtulla, J. C. Milder, T. Kuemmerle, and S. L. Martin. 2013: The elusive pursuit of interdisciplinarity at the human–environment interface. BioScience 63:745–753.

Salazar, M. R., K. Widmer, K. Doiron, and T. K. Lant. 2019: Leader integrative capabilities: A catalyst for effective interdisciplinary teams. Pp. 313–328 *in* K. L. Hall, A. L. Vogel, and R. T. Croyle, eds. Advancing Social and Behavioral Health Research through Cross-Disciplinary Team Science: Principles for Success. Springer, New York.

Schegloff, E. A. 1991: Conversation analysis and socially shared cognition. Pp. 150–171 *in* L. B. Resnick, J. M. Levine, and S. D. Teasley, eds. Perspectives on Socially Shared Cognition. Vol. 150. American Psychological Association, Washington DC.

Shannon, C. E. 1948: A mathematical theory of communication. Bell System Technical Journal 27:379–423.

Stocklmayer, S. M. 2001: The background to effective science communication with the public. Pp. 3–22 *in* S. M. Stocklmayer, R. Goré, and C. R. Bryant, eds. Science Communication in Theory and Practice. Springer, Dordrecht.

Suldovsky, B., B. McGreavy, and L. Lindenfeld. 2018: Evaluating epistemic commitments and science communication practice in transdisciplinary research. Science Communication 40:499–523.

Szabó, Z. G. 2005: Semantics Versus Pragmatics. Clarendon Press, Oxford.

Thompson, J. L. 2009: Building collective communication competence in interdisciplinary research teams. Journal of Applied Communication Research 37:278–297.

Tress, G., B. Tress, and G. Fry. 2004: Clarifying integrative research concepts in landscape ecology. Landscape Ecology 20:479–493.

Tuomela, R. 2007: The Philosophy of Sociality: The Shared Point of View. Oxford University Press, Oxford.

von Wehrden, H., M. H. Guimarães, O. Bina, M. Varanda, D. J. Lang, B. John, F. Gralla, D. Alexander, D. Raines, A. White, and R. J. Lawrence. 2019: Interdisciplinary and transdisciplinary research: Finding the common ground of multi-faceted concepts. Sustainability Science 14:875–888.

Wargad, R., and M. Puri. 2015: A study of cloud computing based model to facilitate collaborative research. International E-Journal of Advances in Education 1:142–148.

Watzlawick, P., J. B. Bavelas, and D. D. Jackson. 1967: Pragmatics of Human Communication. WW Norton, New York.

Wear, D. N. 1999: Challenges to interdisciplinary discourse. Ecosystems 2:299–301.

Whyte, K. P., and R. P. Crease. 2010: Trust, expertise, and the philosophy of science. Synthese 177:411–425.

Wright, J. C., T. Nadelhoffer, T. Perini, A. Langville, M. Echols, and K. Venezia. 2017: The psychological significance of humility. The Journal of Positive Psychology 12:3–12.

Zierhofer, W., and P. Burger. 2007: Disentangling transdisciplinarity: An analysis of knowledge integration in problem-oriented research. Science & Technology Studies 20:51–74.

6 The Power of Philosophy

Chad Gonnerman
Graham Hubbs
Bethany K. Laursen
Anna Malavisi

6.1 WHAT GOOD IS PHILOSOPHY?

There is no shortage of scientists who are skeptical of the power of philosophy. For example, Bill Nye says that "philosophy is important for a while …. But you can start arguing in a circle" (Nye 2016). Neil DeGrasse Tyson claims that philosophy is not "a contributor to our understanding of the natural world" (Hardwick 2014). Stephen Hawking describes philosophy as "dead" (Holt 2012). There are scholars in the humanities who agree. Fish (2011) complains that "the conclusions reached in philosophical disquisitions do not travel. They do not travel into contexts that are not explicitly philosophical (as seminars, academic journals, and conferences are), and they do not even make their way into the non-philosophical lives of those who hold them." One might retort that this is all a matter of what Nathan Ballantyne has called "epistemic trespassing," the inappropriate encroachment of experts in one field into another they do not understand (Ballantyne 2019). However, philosophers have also had reservations about philosophy, at least as it is typically studied and taught in universities today. Frodeman and Briggle (2016) have written that philosophy has "lost its way" and turned into a discipline full of arcane language and obscure debates, detached from—and all-too-often, apparently unconcerned with—both our daily lives and the natural world in which we live. These concerns are not new. Dewey (1917, p. 5) wrote that "Unless professional philosophy can mobilize itself sufficiently to assist in this clarification and redirection of men's thoughts, it is likely to get more and more sidetracked from the main currents of contemporary life."

It is easy to sympathize with these complaints, as it is not uncommon for academic philosophers to lose sight of the forest for the trees. To take but one example, we all should be concerned with whether moral obligations exist and, if they do, what they are. This is one of the basic questions of ethics—it concerns how we should live our lives. Many philosophers think that before we can address such a question, however, we must first determine to what our ethical terms refer and what, if anything, makes sentences with those terms true. If you study these questions, you are a meta-ethicist, and you are likely to have views on whether the fact that normative terms embed in semantically complex sentences causes serious problems for a view about the meaning of moral vocabulary called "expressivism." If you don't study these questions, you may have a hard time following the train of thought in the last sentence, and to the extent that you get that it has something to do with semantics and grammar, you may be baffled as to how it could bear on how we should live our lives. A careful meta-ethicist will be able to explain the connection between our moral judgments, the language we use to make those judgments, and the sorts of linguistic tests we can use to determine whether our analysis of that language is correct or incorrect. A careless meta-ethicist, by contrast, won't care and will talk about compositionality, modality, illocutionary force, and a host of other concepts that don't clearly have anything to do with how to live our lives. These trees may matter if we are to comprehend the forest of ethics. When an outsider watches a furious debate about the details of the leaves, it is unsurprising if they feel some sympathy with the critical views above.

It doesn't have to be this way. Philosophers can be better at explaining how their abstract theorizing bears on concrete problems, and they can spend more time directly addressing these concrete problems. There are plenty of philosophers who acknowledge this—Frodeman and Briggle are examples, as are the contributors to Brister and Frodeman (2020). In fact, the American Philosophical Association, the primary professional organization for philosophers in the United States, has recently asserted that it "values philosophers' participation in the public arena, [which] includes work that engages with contemporary issues as well as work that brings traditional philosophies to non-traditional settings" (APA Communications 2017). For more than a decade, it has supported a formal committee on public philosophy, whose basic charge is "to find and create opportunities to demonstrate the personal value and social usefulness of philosophy" (see www.apaonline.org/members/group.aspx?code=public).

One doesn't have to scratch far beneath the surface of mainstream philosophy to find the work that does this. For example, many philosophers are deeply concerned about how race, class, and gender influence whose views get taken seriously in society. These philosophers use the tools of epistemology, philosophy of language, and ethics to study epistemic injustice, which examines the ways in which people can be disadvantaged when it comes to the production and flow of knowledge, as when a speaker is given less credibility because of a listener's prejudice toward the speaker's (perceived) identity (Longino 1990; Code 1991; Mills 1997; Fricker 2007; Haslanger 2012; Medina 2013; Kidd et al. 2017). Some philosophers of biology investigate the practice of biological research by learning both the biology and the politics of what this science studies (Hull 1988). The same holds for many environmental philosophers. Some philosophers collaborate with scientists, often in the hope of generating philosophical products tailored so that they more likely benefit scientists. This includes work in the biological sciences, climate sciences, neurosciences, and cognitive sciences (Malle and Knobe 1997; Orzack and Sober 2001; Keller 2010; Goes et al. 2011; Schwitzgebel and Cushman 2012; Silva et al. 2013; Strohminger and Nichols 2014; Martin et al. 2016; Mayer et al. 2017; Laplane et al. 2019).

These are just a few of the many ongoing research programs doing what we call *engaged philosophy*. There are also presently several professional organizations committed to this work. The Public Philosophy Network (www. publicphilosophynetwork.net/) is an outgrowth of the APA's Committee on Public Philosophy; it organizes philosophers across North America who are committed to publicly engaged philosophical research and practice. The Humility and Conviction in

Public Life Project (https://humilityandconviction.uconn.edu/) is led by philosophers hoping to improve civil discourse through public forums. The Consortium for Socially Relevant Philosophy of/in Science and Engineering (SRPOISE) (http://srpoise.org/) supports and advances philosophical work related to science and engineering that contributes to public welfare and collective wellbeing. The Joint Caucus of Socially Engaged Philosophers and Historians of Science (JCSEPHS) (https://jointcaucus. philsci.org/) has a similar commitment to publicly meaningful history and philosophy of science, as does the Society for Philosophy of Science in Practice (www. philosophy-science-practice.org/ (see also Ankeny et al. 2011). The list could—and does—go on. Although it may sometimes get lost amidst both the forest and the trees, engaged philosophical efforts such as TDI are a valuable part of current philosophy.

6.2 TDI: AMELIORATIVE ENGAGED PHILOSOPHY

As noted above, engaged philosophy takes a variety of forms. Many of these forms, including those involved in TDI, aim to use philosophy to improve the efficiency and quality of what they study. We call this sort of philosophical practice "ameliorative engaged philosophy." To see what we mean by this, consider TDI's work with interdisciplinary scientific teams, which we discuss in Chapter 2. One of the goals of working with these teams is to help them discover their underlying philosophical beliefs about the practice of science. The goal here is to examine these philosophical underpinnings, not for the sake of critique, but to help the interdisciplinary teams improve their productivity. The members of an interdisciplinary research team can find it difficult to work together due to deeply rooted differences in language and differences in views, attitudes, or commitments concerning the philosophical aspects of science and its objects of inquiry. Engagement in philosophical dialogue, which bolsters mutual understanding of the differences, can help the members of teams with these sorts of difficulties communicate more effectively. This can be an important step for research productivity.

The first several chapters of this book review some of these issues, but let's go a little deeper into the power that philosophy can have to uncover and to mitigate the problems that arise from some of these linguistic, conceptual, and worldview differences. Imagine a research team whose task is to understand invasive species in a specific ecosystem, and which includes a geographer, an ecologist, a lab biologist, and a rural sociologist. The geographer studies the terrain and atmosphere of the ecosystem, the ecologist studies the ecosystem as such, the lab biologist studies the genetics of the invasive species, and the rural sociologist studies the social, political, and economic factors that contribute to the land use of the ecosystem. These scientists are all working on a cross-disciplinary scientific team, but they may have radically different views of what science is and involves despite being members of the same cross-disciplinary scientific team. The lab biologist might say that science requires controlled testing of hypotheses, but the geographer might disagree. The ecologist might say that qualitative data are not real scientific data, but the rural sociologist might say they are. The geographer and lab biologist might believe that good science is value-neutral, while the ecologist and the rural sociologist might believe that it need not be. Such differences in scientists' (implicit) views of the nature of science have recently been explored in empirically informed philosophy of science. For example, Robinson et al. (2019) found that the physical scientists and mathematicians in Toolbox workshops were more likely than the social and behavioral scientists to think that value-neutral scientific research is possible (see also Robinson et al. 2016). Steel et al. (2017) found that scientists who identify as female in Toolbox workshops were more likely than scientists who identify as male to agree (p. 27) that "determining what constitutes acceptable validation of research data is a values issue" (see also Steel et al. 2018). A team with members having different beliefs about the roles of controlled experiments, qualitative data, and values in science does not seem to have a common conception of science. If they lack a common conception, how can they do science together?

A *critical* philosopher of science might examine these differences and conclude that some of these scientists are just wrong. This philosopher might say, "Observational field work and controlled lab experiments may both be called 'science,' but they belong to distinct modes of knowledge-production; ditto the difference between qualitative and quantitative data. The scientific questions we ask are always a reflection of our values, so there is no such thing as value-free science. That solves all of the issues: now, get back to work!" This critical answer does not embody the sort of philosophy TDI does.

What is the philosophy of TDI? Like a lot (indeed, according to some, *all*) of philosophy, the philosophy of TDI is concerned with the definitions of terms and concepts. The critical philosopher above, in response to the disagreement among members of our hypothetical team, might be seen as (implicitly) driven by the question "What does 'science' mean?" The philosophy of TDI may also elicit such questions. As is common in some parts of philosophy, TDI philosophy need not assume that there are correct answers to such definitional questions. Consider our critical philosopher, who doesn't think there is one and only one true definition of "science." In response to the disagreement between the lab biologist and geographer, this philosopher might say, "You could define it to include observational fieldwork or you could exclude fieldwork from the definition." This philosopher thinks it is more important to see observational fieldwork and laboratory experiments as different ways of producing knowledge, with their own philosophically interesting features. It is the distinctions that matter, not the labels. TDI is likewise interested in unearthing the distinctions behind the labels. Finally, like much philosophy, the philosophy of TDI encourages stepping back to take a broad view of what is being discussed. Compare once more the way we have depicted the critical philosopher. They might agree that selectively using data to back one's preferred political view is bad science and go on to note that, nevertheless, values have a role to play in science, if only because it is impossible to fully remove one's values when conducting scientific research or to do science that does not promote some values over others (Elliott 2017).

However, the ameliorative engaged philosophy of TDI does not adopt the critical philosopher's explicitly *critical* goals. TDI sets aside the question of whether there are better or worse answers to these philosophical questions; what matters is figuring out what each scientist on the team *thinks* the right answers are and what each *takes as their reasons* for what they think. TDI uses philosophy, not to get researchers to hold a single view on these sorts of fundamental questions, but to understand their own views, as well as the views of their collaborators, so as to improve the work done by the team. Often scientists (like most people!) haven't explicitly thought about the metaphysical, epistemological, and value-theoretic dimensions of science, even though these dimensions have significant influence on scientific deliberation and decision-making. As a result, implicit and unexamined presuppositions about these dimensions can unconsciously guide one's inferences and judgments. "Know thyself" is a long-standing maxim that most philosophers would endorse. For TDI, as it seeks to help cross-disciplinary science, "know thyself" means knowing your underlying philosophical commitments, which can help to facilitate team communication and functioning. (For more on how TDI facilitates this self-knowledge, see Gonnerman et al. 2015.) Whether there is good reason to change one's view of science is secondary—the primary goal is to get clear to oneself and one's team what one thinks about core issues of science.

Different sorts of self-knowledge are necessary for different sorts of scientific teams. In our invasive-species example above, the team might be able to function well even if the lab biologist persists in believing that geography isn't "really" science. It might be preferable for all members of the team to respect the work of teammates across disciplines, but successful research here might only require effective informational communication, not mutual respect (see Chapter 5). For more highly integrated research projects, disagreement might prove more problematic. When a team works not through piecemeal division of labor but rather as a *collective agent*, substantial agreement can foster cooperation toward a common goal (e.g., List and Pettit 2011; Bratman 2014; Gilbert 2014). The issues here are subtle; timing can matter, as sometimes team improvement requires that agreement not be brought about too quickly (Crowley et al. 2016). No matter what degree of agreement and mutual respect is appropriate for a given research team, effective communication is necessary for its success. Philosophy can help by bringing to the fore fundamental differences in attitudes regarding the philosophical dimensions of science. This improves one's own understanding of science, as well as those of one's collaborators (see Eigenbrode et al. 2007; O'Rourke and Crowley 2013).

Whether and to what extent TDI's ameliorative engaged philosophy actually helps in this way is, of course, an empirical matter—see Chapters 9 and 10 for evidence of its effectiveness. The tools for formulating the problem and for designing possible solutions are, by contrast, an indication of the power of philosophy.

6.3   PUTTING PHILOSOPHY TO WORK

How does one put philosophy to work for research teams? One of the primary techniques of philosophy is conceptual analysis. In one influential mode, conceptual analysis involves taking a term or concept, disambiguating possible different meanings, asking what follows from accepting one meaning instead of another, investigating how competing meanings can produce confused reasoning and conversation, and detecting whether disagreement can be resolved or dissolved by clarifying these competing uses. In Toolbox workshops, we start by using this technique in reverse. The instruments that guide Toolbox workshops include prompts that are designed to be vague and ambiguous, to invite confusion, to promote disagreement. Imagine our invasive-species team addressing the following prompt: "Values have no place in science." We have already considered a geographer and lab biologist agreeing with this statement, and we have considered an ecologist and rural sociologist disagreeing with it. But imagine another team member saying, "Hang on: what do we mean by 'values' and what do we mean by 'no place'? Maybe they have no place in data analysis, but might they have a place in which projects get funded?" This scientist is probing commitments that their teammates are taking for granted. The TDI approach is built on the idea that an excellent way of getting team members to probe their assumptions and presuppositions and to get the other members to put their disagreements on the table is to offer the team prompts that invite multiple interpretations. In part, the goal of the Toolbox conversation is to expose these differences and thereby clarify each other's views. The prompts in TDI instruments are designed to get the differences out in the open.

As discussed in Chapter 5, communication consists of both informational and relational aspects, and the emphasis on conceptual analysis helps mainly with the former, informational aspect. Conceptual analysis reveals the range of concepts that can become inputs into the integration process, and also the range of integrative relations that might be used. In a highly interdependent team, synthesizing these different concepts into shared understanding requires careful attention to the kinds of relationships the team uses to integrate concepts (e.g., subsumption, serialization). Philosophy helps us focus on these connections and construct them logically. This is the power of philosophy to illuminate project-relevant information and enhance our critical thinking skills in the service of supporting substantive project integration.

A Toolbox workshop is also philosophy in action. When philosophers engage in conceptual analysis or in philosophical discourse more generally, their discussions are often conditional and hypothetical: a given term could mean this, or it could mean that, and if it does mean this, then this other thing follows, but if it means that, then still something else follows. This form of discussion does not require ever saying that a given definition is the right definition, that all other definitions are wrong—this form of discussion is exploratory, not dogmatic. It may be that philosophers' willingness not to just "stick to the facts" is what makes the famous scientists listed at the beginning of this chapter frustrated with philosophy. Either way, this anti-dogmatic, open-minded attitude is essential for getting clear on conceptual differences that impede scientific teamwork. Exploring these differences with an open mind is philosophical activity, and so scientists are doing philosophy when they talk about the differences in their responses to Toolbox prompts and when they consider different definitions and reasons for possibly accepting one rather than another.

Toolbox prompts cover not only tricky concepts but also contentious practices the team is facing. A Toolbox prompt might be, "Someone should be a co-author on our papers only if they were on the original grant proposal." This is an opportunity for normative reasoning about the right thing for the team to do. One can imagine the sort of reasons a team member might use to support the claim, such as that only those who contributed to the original grant proposal have earned the right to be a co-author, or to refute it, such as that they should draw on the most relevant expertise whether or not that was originally represented by the team who drafted the proposal. The goal is to get the participants in a workshop to do some philosophy themselves as they discuss the prompts.

This sort of philosophy in action is a powerful way to expose not only hidden assumptions but also hidden ways of reasoning about those assumptions. Getting clear on concepts requires giving reasons for and against alternatives that other people can understand (Laursen 2018). People from different disciplines often appeal to different sorts of reasons for their assertions or beliefs. For example, our rural sociologist above might believe values do have a place in science because they believe that revealing values to respondents helps build rapport with them, and rapport is important for getting informed consent and accurate information. However, the idea of depending on respondent trust has no clear bearing on the lab biologist's work and therefore they may regard it as an inadequate reason to allow values in science—they may even question if rural sociology is a science. Moreover, the lab biologist may have been trained that revealing one's values in a study only creates ethical dilemmas that hamper the science. We know that other aspects of people's backgrounds such as their socioeconomic status, gender, religion, and ethnicity may also influence what they consider to be adequate reasons for work choices. Disciplinary and other diverse backgrounds make different reasons salient to different people. The power of philosophy in dialogue is to make these implicit standards of reasoning explicit.

By doing so, philosophy helps uncover differences among people's opinions. These differences are often anchored in different attitudes toward evidence, success, risk, resource use, and other project-related issues, and differences in these attitudes usually require different conversations. Not only is philosophy good at delineating these different conversations, it is good at helping us have them—both informationally and relationally. From an informational perspective, philosophy invites articulate reasoning with an open mind to alternatives. TDI philosophy focuses this reasoning on concepts and practices that are important to the success of a particular team. The philosophical commitment to suspend analytic judgment for a bit can place everyone metaphorically in the same space so they can have a shared experience of exploring these key issues from each other's perspectives—and recognize and perhaps change their project-related values in the process.

The ideal philosopher's posture—open-minded and charitable—can also help encourage productive relational communication. No one likes to be dominated by another in a conversation; no one likes to be ignored or parodied when real disagreements are on the table. Philosophy asks us to keep an open mind and a charitable spirit when engaging in dialogue with other people about their ideas. When others sense you are actively listening to them, they often feel they can trust you more, which is the foundation of relational communication.

6.4   TDI AND EPISTEMIC JUSTICE

When people come from different backgrounds, it can be hard to decide how to approach group conversation. It can take time to identify these different backgrounds and work out how they might be relevant to the topic of conversation. In some instances, neglecting these differences just to move the conversation forward may feel expedient, but it can also do epistemic damage if participants end up marginalized or silenced in the name of efficiency. Dominant voices might silence dissenters, who may thereafter continue to smother themselves (Dotson 2011). Such epistemic oppression blocks access to knowledge held by the group and is unjust to individuals and their communities. Engaged philosophy like TDI jumps into these messy situations to call a timeout while groups surface and examine their root commitments. This honors each participant and their communities as legitimate knowers. The dialogue further supplies needed intellectual and social resources to the group. Key forms of knowledge become more accessible, and trust begins to take root as each person feels heard, seen, and respected. TDI philosophy is not only an intellectual intervention—it can be a social justice intervention as well, serving to identify moral and epistemic injustice and the production of ignorance (Piso et al. 2016).

As mentioned at the beginning of this chapter, some philosophers use the conceptual tools of epistemology and philosophy of language to study epistemic injustice (Hornsby and Langton 1998; McGowan 2012; Tirrell 2018; Johnson 2019). For many years now, feminists and social philosophers have challenged traditional epistemology for the way it ignores the experiences of some knowers and their capacities as situated knowers. Feminist epistemology has developed in order to provide an alternative epistemology, one that is concerned with context and the relationality of individual lives. Traditional epistemology, because of its apparent abstractness and neutrality, distracts its practitioners from the fact that oppression affects knowledge exchange and production. The analysis of knowledge through the eyes of traditional epistemologists often excludes crucial aspects of our physical and social worlds. From the work of feminist epistemologists, the concept of epistemic injustice—i.e., the idea that justice is not only confined to social, political and cultural factors but also epistemological—has resulted in a growing literature that addresses its meaning and value (Code 1987; Mills 1988; Tuana 2006; Fricker 2007; Dotson 2011, 2012; Anderson 2012; Wanderer 2012; Medina 2013; Pohlhaus 2014; Johnson 2016; Piso et al. 2016).

Critical analyses of epistemology have not been limited to thinking about knowledge but also consider the ways that ignorance plays into epistemically relevant exchanges. So, just as understanding knowledge is important, the way in which we understand ignorance is equally important. Ignorance is thought of as a lack of knowledge, but ignorance can take many forms and have many etiologies. For example, according to Mills (2007), "white" ignorance is a result of structural features in our society that promote and sustain power and domination. Mills is keying in here on a connection between the production of knowledge and broader practices of injustice, which call out for a form of epistemic responsibility that bridges epistemology with the ethico-political (Code 1987).

Feminist epistemology aims to reveal the ethical and political aspects of epistemic conduct. How we constitute knowledge is influenced by our moral and epistemic assumptions, which are tied to socio-cultural and political factors (Fricker 2007). Fricker introduces two forms of epistemic injustice: testimonial and hermeneutical. When a hearer assigns lessened credibility to what someone has said for reasons that are prejudicial and are not relevant to its epistemic status, this can result in testimonial injustice (Fricker 2017). According to Fricker, one-off cases of credibility deficits are epistemic harms, but when the deficits are systematic—as they are, e.g., when women are given less credibility than men on certain topics—they constitute testimonial injustice. Hermeneutical injustice occurs when conceptual resources are unequally distributed, occluding the experience of systematically subordinated agents. If a person lacks the conceptual resources to describe their experience, they will have a hard time understanding it, let alone communicating it to others. If we don't have those resources because we don't attend to the experiences of marginalized groups, the result is hermeneutical injustice.

Toolbox workshops can help reveal such injustice if, in the course of a conversation, participants of marginalized groups become aware of the difficulty they are having articulating their perspective. Workshops can also enhance communication by illuminating epistemic injustice more broadly. The structured dialogue can facilitate the way a team addresses the conditions that can produce distorted forms of knowledge and ignorance. This can be done by including prompts related to testimonial and hermeneutical injustice, e.g., "Certain disciplines are treated with more respect in interdisciplinary collaborations." But it can also take place by facilitators establishing ground rules for dialogue that call attention to power dynamics and implicit biases and encourage participants to observe them while communicating with one another respectfully. (For more on this, see Chapter 8.)

Focusing on epistemic justice can enhance expediency. Consider again the interdisciplinary team described above. Such a focus might lead the team to adopt a paradigm that is the most helpful, is transparent, and is a legitimate "third side" from which to view and integrate all the others. For example, the rural sociologist can champion the rights of marginalized Indigenous and poor residents. This can spark a paradigm shift from a command-and-control resource management paradigm to an adaptive governance paradigm that devolves power to the people interacting most closely with the resource. When enacted well, adaptive governance can be more effective at addressing invasive species, transparently revealing decisions, and giving all parties responsibility and control over the resource.

A philosophical orientation is particularly well suited for achieving these ends. First, philosophy operates at a level of abstraction that is helpful because it does not cast concepts in terms of other disciplines participating in the conversation; that is, it does not beg the question by presupposing a particular disciplinary perspective when unbiased consideration of differences among the perspectives is a workshop goal. For example, philosophy would focus on the root issue of "evidence," which is common between all disciplines, instead of on the distinction between "qualitative" and "quantitative" data. This shared language can help get the conversation started and even enable integrated understanding. Second, engaged philosophy exemplified by the Toolbox workshop makes transparent to participants these abstract issues, as well as any group decision about them, because they are exploring them themselves. Third, engaged philosophy can enhance legitimacy of the conversation and its outcomes for several reasons. One is that the transparency of the process enhances legitimacy. In addition, some people believe that legitimate ways of thinking only come from established disciplines and thus find emerging fields to be suspect. For them, philosophy certainly qualifies as an established discipline. Last, others believe legitimate lenses are created only by the participants. In engaged philosophy such as that practiced by TDI, every conversation is unique. In summary, when a team engages in a Toolbox workshop, philosophy gives it informational *and* relational resources that can enable the team to perform both more efficiently and more justly.

The informational and relational resources of philosophy (see also Chapter 5) can boost team performance because they enable the right kind of coordinated action aiming at the team's goals. Some teams only need awareness of each other's perspectives and practices, and Toolbox workshops are great for this. But highly interdependent teams not only need awareness; they also need coordination. Sometimes they may generate coordination during a Toolbox workshop. Either way, in a Toolbox workshop, a team practices philosophical skills like conceptual abstraction and value analysis that enable them to have *future* conversations in which they work through details and adapt to changes in their project (Laursen 2018). They can continue to enhance their shared understanding over time through such practice, not just in the workshop. This is the ongoing power of philosophy.

6.5 ACKNOWLEDGMENTS

We would like to acknowledge engaged philosophers, particularly those in Departments of Philosophy at Michigan State University and the University of Idaho, who help us understand what philosophy can do for today's world.

6.6 LITERATURE CITED

Anderson, E. 2012: Epistemic justice as a virtue of social institutions. Social Epistemology 26:163–173.

Ankeny, R., H. Chang, M. Boumans, and M. Boon. 2011: Introduction: Philosophy of science in practice. European Journal for Philosophy of Science 1:303–307.

APA Communications. 2017: *APA Statement on Valuing Public Philosophy*.

Ballantyne, N. 2019: Epistemic trespassing. Mind 128:367–395.

Bratman, M. E. 2014: Shared Agency: A Planning Theory of Acting Together. Oxford University Press, Oxford.

Brister, E., and R. Frodeman. 2020: A Guide to Field Philosophy: Case Studies and Practical Strategies. Routledge, New York.

Code, L. 1987: Epistemic Responsibility. University Press of New England, for Brown University Press, Hanover and London.

———. 1991: What Can She Know?: Feminist Theory and the Construction of Knowledge. Cornell University Press, Ithaca, NY.

Crowley, S. J., C. Gonnerman, and M. O'Rourke. 2016: Cross-disciplinary research as a platform for philosophical research. Journal of the American Philosophical Association 2:344–363.

Dewey, J. 1917: The need for a recovery of philosophy. Pp. 3–69 *in* J. Dewey, A. W. Moore, H. C. Brown, G. H. Mead, B. H. Bode, H. W. Stuart, J. H. Tufts, and H. M. Kallen, eds. Creative Intelligence: Essays in the Pragmatic Attitude. Henry Holt, New York.

Dotson, K. 2011: Tracking epistemic violence, tracking practices of silencing. Hypatia 26:236–257.

———. 2012: A cautionary tale: On limiting epistemic oppression. Frontiers: A Journal of Women Studies 33:24–47.

Eigenbrode, S. D., M. O'Rourke, J. D. Wulfhorst, D. M. Althoff, C. S. Goldberg, K. Merrill, W. Morse, M. Nielsen-Pincus, J. Stephens, L. Winowiecki, and N. A. Bosque-Pérez. 2007: Employing philosophical dialogue in collaborative science. BioScience 57:55–64.

Elliott, K. C. 2017: A Tapestry of Values: An Introduction to Values in Science. Oxford University Press.

Fish, S. 2011: *Does Philosophy Matter?* Downloaded from https://opinionator.blogs.nytimes.com/2011/08/01/does-philosophy-matter/.

Fricker, M. 2007: Epistemic Injustice Power and the Ethics of Knowing. Oxford University Press, Oxford.

———. 2017: Evolving concepts of epistemic injustice. Pp. 53–60 *in* I. J. Kidd, J. Medina, and G. Pohlhaus Jr, eds. The Routledge Handbook of Epistemic Injustice. Routledge, London.

Frodeman, R., and A. Briggle. 2016: *When Philosophy Lost its Way*. Downloaded from https://opinionator.blogs.nytimes.com/2016/01/11/when-philosophy-lost-its-way/.

Gilbert, M. 2014: Joint Commitment: How We Make the Social World. Oxford University Press, Oxford.

Goes, M., N. Tuana, and K. Keller. 2011: The economics (or lack thereof) of aerosol geoengineering. Climatic Change 109:719–744.

Gonnerman, C., M. O'Rourke, S. J. Crowley, and T. E. Hall. 2015: Discovering philosophical assumptions that guide action research: The reflexive toolbox approach. Pp. 673–680 *in* H. B. Huang and P. Reason, eds. The SAGE Handbook of Action Research, 3rd edition. SAGE, Thousand Oaks, CA.

Hardwick, C. 2014: *Neil deGrasse Tyson Returns Again*. Downloaded from http://nerdist.nerdistind.libsynpro.com/neil-degrasse-tyson-returns-again.

Haslanger, S. A. 2012: Resisting Reality: Social Construction and Social Critique. Oxford University Press, Oxford.

Holt, J. 2012: Physicists, stop the churlishness. New York Times: Vol, 161, Issue 55798, p. 12.

Hornsby, J., and R. Langton. 1998: Free speech and illocution. Legal Theory 4:21–37.

Hull, D. L. 1988: Science as a Process. University of Chicago Press, Chicago.

Johnson, C. R. 2016: If you don't have anything nice to say, come sit by me: Gossip as epistemic good and evil. Social Theory and Practice 42:304–317.

———. 2019: Investigating illocutionary monism. Synthese 196:1151–1165.

Keller, D. R. 2010: Environmental Ethics: The Big Questions. John Wiley & Sons, New York.

Kidd, I. J., J. Medina, and G. Pohlhaus Jr. 2017: Routledge Handbook of Epistemic Injustice. Routledge, New York.

Laplane, L., P. Mantovani, R. Adolphs, H. Chang, A. Mantovani, M. McFall-Ngai, C. Rovelli, E. Sober, and T. Pradeu. 2019: Opinion: Why science needs philosophy. Proceedings of the National Academy of Sciences 116:3948–3952.

Laursen, B. K. 2018: What is collaborative, interdisciplinary reasoning? The heart of interdisciplinary team research. Informing Science 21:75–106.

List, C., and P. Pettit. 2011: Group Agency: The Possibility, Design, and Status of Corporate Agents. Oxford University Press, Oxford.

Longino, H. E. 1990: Science as Social Knowledge: Values and Objectivity in Scientific Inquiry. Princeton University Press, Princeton, NJ.

Malle, B. F., and J. Knobe. 1997: The folk concept of intentionality. Journal of Experimental Social Psychology 33:101–121.

Martin, C. B., R. A. Cowell, P. L. Gribble, J. Wright, and S. Köhler. 2016: Distributed category-specific recognition-memory signals in human perirhinal cortex. Hippocampus 26:423–436.

Mayer, L. A., K. Loa, B. Cwik, N. Tuana, K. Keller, C. Gonnerman, A. M. Parker, and R. J. Lempert. 2017: Understanding scientists' computational modeling decisions about climate risk management strategies using values-informed mental models. Global Environmental Change 42:107–116.

McGowan, M. K. 2012: On 'whites only' signs and racist hate speech: Verbal acts of racial discrimination. Pp. 121–147 *in* I. Maitra and M. K. McGowan, eds. Speech and Harm: Controversies Over Free Speech. Oxford University Press, Oxford.

Medina, J. 2013: The Epistemology of Resistance: Gender and Racial Oppression, Epistemic Injustice, and Resistant Imaginations. Oxford University Press, Oxford.

Mills, C. W. 1988: Alternative epistemologies. Social Theory and Practice 14:237–263.

———. 1997: The Racial Contract. Cornell University Press, Ithaca, NY.

———. 2007: White ignorance. Pp. 26–31 *in* S. Sullivan and N. Tuana, eds. Race and Epistemologies of Ignorance. State University of New York Press, Albany.

Nye, B. 2016: *Hey Bill Nye, Does Science Have All the Answers or Should We Do Philosophy Too?* Downloaded from www.youtube.com/watch?v=ROe28Ma_tYM.

O'Rourke, M., and S. J. Crowley. 2013: Philosophical intervention and cross-disciplinary science: The story of the Toolbox Project. Synthese 190:1937–1954.

Orzack, S. H., and E. Sober. 2001: Adaptationism and Optimality. Cambridge University Press, Cambridge.

Piso, Z., E. Sertler, A. Malavisi, K. Marable, E. Jensen, C. Gonnerman, and M. O'Rourke. 2016: The production and reinforcement of ignorance in collaborative interdisciplinary research. Social Epistemology 30:643–664.

Pohlhaus Jr, G. 2014: Discerning the primary epistemic harm in cases of testimonial injustice. Social Epistemology 28:99–114.

Robinson, B., S. E. Vasko, C. Gonnerman, M. Christen, M. O'Rourke, and D. Steel. 2016: Human values and the value of humanities in interdisciplinary research. Cogent Arts & Humanities 3:1123080.

Robinson, B., C. Gonnerman, and M. O'Rourke. 2019: Experimental philosophy of science and philosophical differences across the sciences. Philosophy of Science 86(3):551–576.

Schwitzgebel, E., and F. Cushman. 2012: Expertise in moral reasoning? Order effects on moral judgment in professional philosophers and non-philosophers. Mind & Language 27:135–153.

Silva, A. J., A. Landreth, and J. Bickle. 2013: Engineering the Next Revolution in Neuroscience: The New Science of Experiment Planning. Oxford University Press, Oxford.

Steel, D., C. Gonnerman, and M. O'Rourke. 2017: Scientists' attitudes on science and values: Case studies and survey methods in philosophy of science. Studies in History and Philosophy of Science Part A 63:22–30.

Steel, D., C. Gonnerman, A. M. McCright, and I. Bavli. 2018: Gender and scientists' views about the value-free ideal. Perspectives on Science 26:619–657.

Strohminger, N., and S. Nichols. 2014: The essential moral self. Cognition 131:159–171.

Tirrell, L. 2018: Authority and gender: Flipping the F-switch. Feminist Philosophy Quarterly 4.

Tuana, N. 2006: The speculum of ignorance: The women's health movement and epistemologies of ignorance. Hypatia 21:1–19.

Wanderer, J. 2012: Addressing testimonial injustice: Being ignored and being rejected. The Philosophical Quarterly 62:148–169.

7 The Power of Dialogue

Michael O'Rourke
Troy E. Hall
Bethany K. Laursen

7.1 INTRODUCTION

We begin by summarizing our argument for the use of dialogue in the Toolbox approach. In Chapter 4, we described challenges that confront people who engage in cross-disciplinary collaboration. Among these challenges, those associated with communication and integration are especially vexing, given that unacknowledged differences in culture and epistemology across disciplines and professions can engender unreasonable agreements and unreasonable disagreements (Crowley et al. 2016). Given their importance to successful cross-disciplinary collaboration, communication and integration received special consideration in Chapter 5.

As we described in Chapter 6, the Toolbox approach assumes (with good evidence—see Chapters 9 and 10) that philosophical analysis can expose important conceptual dimensions along which unacknowledged differences reside and make those dimensions accessible for consideration by members of a cross-disciplinary team. By using philosophy to structure collective reflection on the similarities and differences among the participants, the Toolbox workshop enhances mutual understanding and facilitates more effective communication and integration. In workshops, we realize this sort of collective reflection in *dialogue*, which is a mode of communication that has well-known collective benefits. This chapter addresses the step in our argument from philosophical analysis to collective reflection by providing an overview of dialogue as we understand it in TDI.

Our belief is that a cross-disciplinary team can ameliorate communication or integration issues by talking about them. However, not just any form of talk will do. We argue that dialogue is a form of talk that is essential in orienting participants to differences that can obstruct communication and integration.

Toolbox workshops strive to generate *productive* dialogues, which we define as dialogues that move toward mutual understanding. Consider the dialogue in Table 7.1 from a Toolbox workshop in which participants discussed the nature of replication, how it connects with experimentation, and the relationship of replication and experimentation with science.

The dialogue in this transcript excerpt contains expressions of disciplinary differences, elaborations on previous speaking turns, questions designed to focus the conversation, requests for information, and detailed illustrations. All of these signal

TABLE 7.1

Exchange Drawn from a Toolbox Workshop Involving Two Chemists, an Ecologist, an Economist, and an Engineer

Ecologist	So do you believe that if you can't do the experiment, it's not science?
Chemist 1	Well, it depends what you mean by "experiment." I would say what *you* do in looking for patterns has an element of experiment in it. You go out, you observe; you look for patterns. It's an observational experiment. So experimentation and observation are two pieces of science.
Economist	So it doesn't require replication, or it does require replication?
Chemist 1	Oh, in the science that I do, replication is important.
Chemist 2	[But] Sometimes you just can't replicate.
Chemist 1	I'd have to be given a "for instance," probably, to be able to test that.
Engineer	Let's look at evidence of climate change, going back and looking at ice cores or other information. I mean, the replication is not necessarily actually performing experiments using comparable, sort of, spatio-temporal sets of data .... I was watching this Nova show on super volcanoes and it pointed to three different events that were occurring. Three different measurements that were made of post-data, and they all came together ... it was completely independent work, but replication .... It's not true replication, but I mean, it's sort of validation in the sense that all three were coincident on exactly the same time period when this giant volcano went off.
Economist	So [let's say] we want to understand and prevent the recurrence of the Rwandan genocide. That's a unique event. You certainly don't want to replicate it, right? So ... can you do science on—
Chemist 1	That's right, and there are some things where you wouldn't want to set up the experiment for moral reasons .... In fact, even the first time is not an experiment. It's an event or a series of events that has an outcome, and you're just observing it. So I don't think you can set that up as an ... it's not set up originally as an experiment. There's no controls, as near as I know; you're just observing it.
Economist	But can you do science on it, is the question.
Chemist 1	[overlap] I don't know whether they call it science, but ... it's social .... I think that social science and history and whatever, I think all those pieces have a component. It's not like physical [science].
Economist	So the Rwandan genocide: unique event; purely observational. As we try to explain what happened, we can test and falsify certain hypotheses. So we can sort of rule out some explanations as we go ... because [let's imagine that] this exact same constellation of activities occurred [in another case], and there was no genocide. So we can verify that this combination does not deterministically cause genocide. So we cannot explain the event, but we can rule out causal factors or causal combinations of factors .... So does that then make it science?
Engineer	I think it's science. I think it does. Social science, for sure.
Chemist 1	I think I wouldn't call it science.

the deep listening, meaning co-construction, mutual engagement, and constructive argumentation that mark productive dialogue, focusing attention and creating synergy that transforms a group of "I"s into a "we" (cf. Tuomela 2007).

Our primary goal in this chapter is to illustrate how dialogue can enhance mutual understanding and integration of perspectives. In previous chapters, we showed how Toolbox workshops use philosophy to structure and focus dialogue so that it can support these goals, but dialogue itself is not monolithic—it comes in different forms, some of which are central to productive exchanges that occur in Toolbox workshops. As a consequence, we adopt an ecumenical approach that theorizes a number of different conversational types as dialogue.

In what follows, after briefly describing what dialogue is *not*, we will describe what it *is*. We then relate our definition of "dialogue" to a range of theories from philosophy, social and organizational psychology, and education. This elaboration builds upon a distinction we draw between *dialogue as argumentation* and *dialogue as association*. We also consider the benefits of dialogue, both for individuals and for groups and teams. We close by discussing how dialogue is fostered in Toolbox workshops.

7.2   THE NATURE OF DIALOGUE

7.2.1   WHAT DIALOGUE IS NOT

To begin, we follow Aloni (2013) in focusing on what dialogue is *not*. For starters, one can distinguish between a monologue, which is delivered by an individual, and a dialogue, which requires at least two participants to speak to each other. A borderline situation occurs when two or more participants contribute speaking turns successively but independently, i.e., without modifying their contributions based on the contributions of the other (Cherry 1957; Traxler 2012). This might happen in a heated conversation in which speakers simply assert their different positions. We agree with those who argue that such a conversation is essentially "chunks of monologue stuck together" (Pickering and Garrod 2004, p. 170), and we do not treat such exchanges as dialogue.

Plural participation is necessary but not sufficient for dialogue. Other conditions may be sufficient but not necessary. For example, some scholars believe that dialogue "always involves significant content," is "always respectful and open," and "evinces ... a spirit of democracy" (Aloni 2013, pp. 1071–1072). While the necessary condition of plural participation alone is too permissive—letting too many things count as dialogue—the latter set of sufficient conditions is overly restrictive and rules out important exchanges that should qualify as dialogue. While we will argue that dialogue is grounded in the ability of interlocutors to pursue a collective goal, we disagree with Aloni that dialogue will "always" involve "significant content," respect, and "a spirit of democracy." We want to include as dialogue verbal exchanges involving speakers who may be suspicious of one another, perhaps don't respect one another, and whose participation may not be overly democratic. We understand dialogue as a medium that can improve the relational connection of interlocutors, so we don't want to assume open, respectful relationships as a precondition. This is not to say that dialogue can begin out of nothing; we agree with Campolo and Turner (2002) and Campolo (2005) that participants must possess some common ground to begin a dialogue. However, conversational partners need not occupy this common ground in a way that qualifies as "respectful and open" or evinces a full "spirit of democracy."

7.2.2 Defining "Dialogue"

Conversation comes in a variety of forms, from superficial and distracted serial monologue to deeply engaged dialogue. For Pickering and Garrod (2004), monologue and dialogue are ends of a continuum based on the type and amount of interaction that takes place between interlocutors. Other things being equal, the more interaction, the more dialogical the exchange. Bangerter and Clark (2003) help us understand the relationship between dialogue and interaction, arguing that (p. 196) "people create dialogue *in the service of* the basic joint activities they are engaged in." Unlike serial monologue, dialogue is used to accomplish some joint aim that all participants recognize and endorse—even if they don't respect each other. So the more two people interact in pursuit of a joint aim, the more need they will have for dialogue as their medium of verbal exchange. This is emphasized by Bangerter and Clark (2003), who hold that the primary value of dialogue is instrumental, its purpose being to support the accomplishment of other joint activities (cf. Tsoukas 2009, pp. 944–945).

The interaction in dialogue is between two or more participants, each of whom is both a speaker and a listener and so is a reciprocal partner in the endeavor (Floyd 2010). Listening ensures connectivity among the ideas that emerge across speaking turns. Participants must listen to each other, process the others' meanings, and shape their own statements to acknowledge and reflect this processing if they are to build collaboratively on each other's contributions (Bakhtin 1981, p. 282; Kuhn 2015, p. 50). That is, dialogue requires collaborative meaning-making (see Chapter 5). Pickering and Garrod (2004, pp. 184–185) note that each participant is always a listener, even when they speak, because speakers are constantly self-monitoring, making adjustments in light of the signals (such as facial expressions) they receive from their listeners. Traxler (2012, p. 320) builds on this and notes that "participants in dialogue simultaneously play the roles of speaker and listener—we monitor our own speech to make sure it comes out right and we plan what we are going to say next while we listen to other people speak." Each participant attends to the way in which conversation unfolds and plans where it should go next. The speaking turns are just the surface activity of the participants, who are also always engaged in more subterranean activity aimed at achieving their joint goal.

Interaction between participants (each of whom is a speaker/listener) and the joint communicative activity this interaction constitutes are key elements in our definition of "dialogue." For our purposes in TDI, it makes sense to embrace a broader conception of dialogue that is grounded in these elements. Such a conception emphasizes the "structural" elements of dialogue, i.e., the basic elements of a dialogue and their interrelationships, such as multiple participants, different speaking turns, and a goal that brings together separate actions into a joint activity. These elements are integrated into the definition of "dialogue" we adopt in this chapter and in TDI more generally, which we draw from Tsoukas (2009, p. 943):

> A dialogue is a joint activity between at least two speech partners, in which a turn-taking sequence of verbal messages is exchanged between them, aiming to fulfill a collective goal.

This structural definition accommodates interaction and joint activity and also the connection of speaker to listener. It also reflects the conception of communication as the co-construction of meaning in pursuit of a goal (see Chapter 5) by emphasizing the collaborative, goal-directed nature of dialogical interaction.

Despite our disagreement with Aloni on some aspects of dialogue, our commitment to this structural definition supports our agreement with him that an adequate account of dialogue needs to be *normative*. His account, for example, specifies norms such as respectfulness and openness which support evaluation of dialogical episodes as *better* or *worse*. We also endorse a conception of dialogue that allows us to distinguish better instances of dialogue (e.g., more productive, more engaging) from worse ones. For us this distinction is drawn in terms of dialogue's ability to support achievement of a collective goal, which will be grounded in mutual understanding. Such a conception can be used to distinguish between Toolbox "wins" (i.e., moments of dialogue in Toolbox workshops that enhance mutual understanding) and Toolbox "fails" (i.e., moments of dialogue in Toolbox workshops that undermine mutual understanding or that have no effect on it).

We can illustrate our definition and the workshop goal of mutual understanding with the help of the excerpt quoted above. The excerpt records an exchange among five speakers who engage in a turn-taking sequence of verbal messages. Although the exchange covers a variety of themes, the speakers aim to address the initial question about the relationship between experimentation and science. As such, what unites these speakers is the goal of contributing to an understanding of what counts as science. The elements of our definition are in place in the excerpt—namely, multiple speakers, conversational turn-taking, pursuit of a collective goal—and so it qualifies as an example of what we take to be dialogue. In addition, there is reason to believe the goal is achieved (at least in part) and so it qualifies as *productive* dialogue.

Thus, by emphasizing pursuit of a collective goal, our structural definition excludes serial monologue, but it does not require so much as to exclude other, more robust conceptions of dialogue found in the literature. In developing our account of dialogue around this definition, we will consider the various forms of dialogue from two different perspectives: the perspective of *argumentation* and the perspective of human *association*. These perspectives shed light on two major kinds of collective goals participants often try to achieve through dialogue. The first perspective brings attention to how dialogue enables people to reason together, while the second perspective brings attention to how dialogue enables people to establish closer associations.

The argumentation and association perspectives on dialogue reveal two ways in which it adds value to Toolbox workshops. Dialogue enhances Toolbox workshops by making the informational and relational aspects of team communication more salient for participating cross-disciplinary teams, and the two perspectives foreground

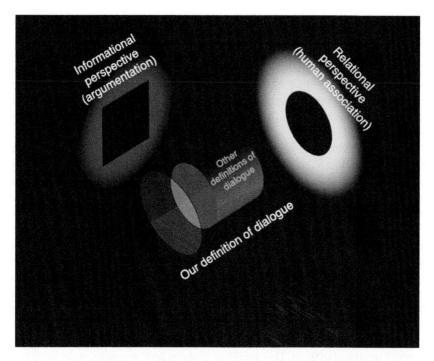

FIGURE 7.1 Our definition of "dialogue" encompasses other definitions, e.g., Aloni's, by requiring the satisfaction of fewer necessary conditions. At the same time, however dialogue is defined it can be viewed from two perspectives: argumentation and human association, which respectively track informational and relational aspects. (Background image by Jean-Christophe Benoist, CC-BY 2.5.)

how dialogue does this. By showing how dialogue makes topical content available for joint consideration (i.e., how dialogue *thematizes* its content, to borrow from Habermas 1984), the argumentation perspective emphasizes the *informational* dimension of communication. The human association perspective emphasizes the potential of dialogue to enrich the *sociopragmatic* and, more specifically, the relational dimensions of communication (see Chapter 5 and Keyton 1999). Thus, these two perspectives highlight attributes of conversational exchanges that often add value to Toolbox workshops for the cross-disciplinary teams that participate in them. Figure 7.1 illustrates our approach.

7.2.3 An Argumentation Perspective on Dialogue

The notion of *argumentation* has provided an important lens for thinking about dialogue. In fields as diverse as education, argumentation theory, and deliberative democracy, conceiving of dialogue in connection with argumentation, and argumentation in connection with dialogue, has advanced both theoretical and practical goals. This perspective assumes that argumentation can be understood as a collaborative or social process involving individuals who critique and defend different positions. Nussbaum

(2008), for example, defines the process of argumentation as one in which individuals engage each other in dialogue where they construct and critique arguments understood as series of propositions containing a conclusion that is derived from premises. This way of thinking about arguments is illustrated by the first contribution from Chemist 1 in the excerpt above, in which they argue that what the Ecologist does in "looking for patterns" implies that "experimentation and observation are two pieces of science." That is, Chemist 1 constructs an argument consisting of premises about the Ecologist's practice and about science to support a conclusion that science has two parts.

Thinking about dialogue in connection with argumentation is useful in part because it reveals how information is exchanged, modified, and adopted through reasoning together. Argumentation highlights the relevance of and reasoning behind opposing positions (Kuhn and Udell 2007). Such critical thinking can encourage "constructive conflict" in which differences are negotiated and resolved by team members (Van Den Bossche et al. 2006). In the excerpt, we see examples of such conflict, as when Chemist 1 requests an "instance" to illustrate a claim, and at the end when the Engineer and Chemist 1 respond in opposite but thoughtful ways to the question from the Economist.

Habermas's conception of dialogue as a vehicle for argumentation is an important foundation for fields of scholarship that consider dialogue in connection with argumentation, three of which are education, argumentation theory, and deliberative democracy, considered below. According to Habermas (1984), argumentation is discourse in which participants put forward "validity claims" (i.e., propositions) and supply reasons either supportive or critical of them. This definition explains how argument, understood as the inference of conclusions from premises, is manifested in social situations: conclusions are "validity claims" and the premises supporting these conclusions can be evaluated by participants in dialogue. For Habermas (p. 10), deliberative discourse that engages interlocutors in the evaluation of a controversial validity claim constitutes rational "communicative action." Communicative action involves participants who coordinate their actions based on mutual understandings they develop through argumentation (Habermas 1996). The Toolbox approach aims to harness dialogue to enhance mutual understanding, and Habermas's approach to argumentation provides insight into at least one dialogical mechanism that can bring about shared understanding, viz., the mutual evaluation of reasons for claims (cf. Laursen 2018a).

Kuhn (2015) similarly describes the value of dialogue for interpersonal argumentation in education. Contrasting spoken, collaborative exchanges between students (i.e., dialogue) with written essays (essentially monologues), Kuhn notes that productive arguments require opposing claims to be in play and that dialogue has a structure that illuminates these claims. She argues that in collaborative discourse (what we call *dialogue*) it is participants' expressions of and challenges to opposing ideas that make the discourse productive. She claims (p. 50) that dialogues "demand attention to the other" and that they engage arguers more "deeply and authentically" than essays. In exchanges where participants defend different sides in an effort to reach agreement, productive dialogue requires that the other side is heard and that one recognizes that one's own side is "contestable." In the context of educational theory, then, thinking of dialogue as a vehicle for argumentation underscores both the need for participants to pay attention to each other and the importance of a set of exchanges that must be understood collectively as providing grounds for believing one position or another. So understood, dialogue is a major way that information is exchanged, modified, and adopted by team members.

In presenting dialogue as a vehicle for informationally valuable argumentation, Kuhn (2015) makes a contribution to education theory that explicitly draws on the work of Douglas Walton (cf. Kuhn and Moore 2015). In his seminal contribution to argumentation theory, Walton (1999) defines "dialogue" explicitly in terms of argumentation. Acknowledging the influence of Grice's (1989) work on conversational maxims, Walton (p. 77) takes dialogue to be:

> a goal-directed conventional framework in which two speech partners reason together in an orderly way, according to the rules of politeness, or normal expectations of cooperative argument appropriate for the type of exchange they are engaged in.

Walton (2010) notes that this framework includes an opening stage, an argumentative stage, and a closing stage. These ways of thinking about dialogue are consistent with the structural definition we adopt above, although they restrict it in various ways, e.g., by suggesting that the "rules of politeness" apply, or that dialogues have three stages. As such, this definition of *dialogue* is one of the more limited ones accommodated by the definition we've adopted.

Understood as Walton has characterized them, dialogues differ according to the initial situation, the goals involved (both individual and collective), and the turn-taking moves and methods of argumentation employed by participants. Table 7.2 summarizes seven types of dialogue that Walton has developed in detail. All of these

---

TABLE 7.2
Seven Basic Types of Dialogue, Taken from Walton (2010, p. 1)

Type of Dialogue	Initial Situation	Participant's Goal	Goal of Dialogue
Persuasion	Conflict of opinions	Persuade other party	Resolve or clarify issue
Inquiry	Need to have proof	Find and verify evidence	Prove (disprove) hypothesis
Discovery	Need to find an explanation of facts	Find and defend a suitable hypothesis	Choose best hypothesis for testing
Negotiation	Conflict of interests	Get what you most want	Reasonable Settlement both can live with
Information seeking	Need information	Acquire or give information	Exchange information
Deliberation	Dilemma or practical choice	Coordinate goals and actions	Decide best available course of action
Eristic	Personal conflict	Verbally hit out at opponent	Reveal deeper basis of conflict

are modes of exchange that enhance mutual understanding and can occur in Toolbox workshops—even what Walton calls "eristic" dialogue, which is grounded in personal conflict but can nevertheless be used to accomplish a shared goal of illuminating core values and assumptions because it reveals the "deeper basis of conflict."

In addition to education and argumentation theory, deliberative democracy is a third field in which dialogue and argumentation are interrelated. This field considers *deliberation* to be conversation in which reasons are exchanged in an open, inclusive, authentic, and respectful pursuit of a "rationally motivated consensus" (Goodin 2005, p. 185). Sprain and Black (2018) emphasize (p. 338) that deliberation is an "interactional accomplishment" that can include disagreement alongside inclusive, respectful, reason-giving exchanges; it is (p. 341) "about awareness of self, other, and issue" generated through an inquiry that involves the provision and evaluation of reasons from multiple perspectives. So understood, deliberative conversation is argumentative discourse and it qualifies as dialogue under the structural definition we introduced above. It is a joint activity that aims to achieve a collective goal via turn-taking and conversational processes.

Argumentation as a lens for understanding dialogue helpfully focuses our attention on the ways in which dialogical interaction among speakers can integrate different perspectives into a coherent combination, as we see in Toolbox dialogues. Each of the three domains in which dialogue is understood in connection with argument—namely, education, argumentation theory, and deliberative democracy—emphasizes the value of dialogical interaction as a mechanism for introducing and evaluating competing reasons for claims.

7.2.4 AN ASSOCIATION PERSPECTIVE ON DIALOGUE

Examining dialogue from the perspective of human association emphasizes the affective characteristics that bring people into closer association when they engage dialogically. Dialogue as association highlights the ways in which people can engage in each other's worlds through conversation (Bakhtin 1981). Whereas the argumentation perspective just discussed considers how dialogue involves the exchange of propositions understood as reasons to reach conclusions, this perspective emphasizes the power of dialogue to transform and improve human relationships, especially when dialogue includes "deep listening, perspective taking, respect, and a sense of genuineness or honesty" (Sprain and Black 2018, p. 341). As Figure 7.1 shows, dialogue can be observed from both perspectives simultaneously.

Aloni's (2013) narrower conception of dialogue, which we introduced earlier, emphasizes such elements of human association. We do not require these features for an exchange to count as dialogue, but we recognize that many dialogues do possess these features when they aim for relational growth. Aloni observes that many forms of conversation—such as small talk, lecture, or commands with responses—often lack these features. His definition of dialogue limits it to exchanges that significantly deepen the relationships of those who participate (p. 1072):

> Dialogue is a conversation in which those involved are attentive to one another and exhibit a mutual interest on the basis of their shared humanity and individual

personalities; out of a shared sense of trust, respect and openness, they jointly advance to a more comprehensive understanding of themselves, others and the circumstances they share.

As we have noted above, his account of dialogue is more limiting than ours, although a rich dialogue that accomplishes what Aloni describes here would qualify as a dialogue under our definition since there is a collective pursuit of a goal through conversational turn-taking.

A similar view of dialogue that emphasizes human association is offered by Johannesen (1971), who takes it to exhibit qualities such as (p. 375) "mutuality, openheartedness, directness, honesty, spontaneity, frankness, lack of pretense, nonmanipulative intent, communion, intensity, and love in the sense of responsibility of one human for another." This explicitly relational conception of dialogue "refers ... to an 'attitude' toward or spirit of communication" (Floyd 2010, p. 128; cf. Johannesen 2002, p. 58). So understood, dialogue is a mode of communication that both represents and contributes to deep human connection.

Buber and Bohm are two important contributors to the theory of dialogue who emphasize the importance of human association. In his classic work, *I and You*, Buber (1996) develops a conception of dialogue that is grounded in his reflections on the relationships between two beings as the self and the other. For Buber, all human interaction is understood in terms of communicative exchanges, such as dialogue, but the nature of that communicative exchange can differ. One can relate to the other as an *I* to an *It*, which obtains between discrete beings and rests on description, analysis, experience, and use. A more primitive and profound relation obtains between the *I* and the *You*, which is an "unmediated" relation between beings who experience one another reciprocally or with "tenderness" (p. 79). The *I-You* relationship involves a spiritual or "cosmic" bond between beings that are neither part of one another nor outside of one another. For Buber, true dialogue can only obtain between an *I* and its *You* as they interact with reciprocity. It involves an "awakening of other-awareness that occurs in, and through a moment of meeting" (Cissna and Anderson 2002, p. 174; see also Black 2008).

To Bohm (1996), dialogue enables a sharing of consciousness and *thinking together* that renders our meanings coherent (cf. Isaacs 1999). Bohm's "vision of dialogue is one in which people pick up thoughts latent in the group and develop them collectively" (p. 45), with each person "partaking of the whole meaning of the group and also taking part in it" (p. 31). This generates a kind of *participatory consciousness* marked by collective reflection on the impulses and assumptions that are made by members of the dialogue group. Gaining this collective achievement requires "close connection" and "mutual participation" (p. 37), with interpersonal fellowship being a strong motivation for dialogue.

In our dual-perspective approach to dialogue, Bohm's view is especially complex, combining elements that reflect both the association and argumentation perspectives. In addition to the associational elements mentioned in the previous paragraph, Bohm understands dialogue to help us identify obstacles to thinking and reveal tacit assumptions, which are important argumentative functions of dialogue. But, unlike most argumentation perspectives, Bohm asserts that coherent meanings arise not from reasoning together but from more relational associations among participants than our definition requires. For Bohm (p. 30), "the object of dialogue is not to analyze things or to win an argument, or to exchange opinions. Rather, it is to suspend your opinions and to look at the opinions—to listen to everybody's opinions, to suspend them, and to see what all that means." For us, by contrast, dialogue need not require suspension of opinions. Two conversational partners could engage in dialogue without such deep, personal connections by conversationally pursuing a collective goal that might be joint analysis of a topic or even just the exchange of opinions.

Dialogue that exhibits argumentation and dialogue that exhibits association are both found in Toolbox workshops, and each type can advance the participants toward a greater degree of mutual understanding. Toolbox workshops ideally function in such a way that participants teach their perspectives to others and also learn about the perspectives of others. These interactions can have both informational and relational aspects as elaborated by the argumentation and association perspectives on dialogue. In the next section, we provide evidence concerning the benefits of dialogue, showing more clearly how Toolbox dialogues aid participants through these informational and relational aspects.

7.3   THE BENEFITS OF DIALOGUE

Productive dialogue has many potential benefits. These include the development of individual and group awareness, reasoning skills, and mindsets that can be useful within a given research project and in scientific practice more broadly. In this section, we first discuss key benefits for individuals that arise from engaging in dialogue. We then move on to explore the broader benefits that may accrue to groups or teams, particularly cross-disciplinary research teams. In doing so, we weave together empirical research gleaned from various disciplines, including education, team science, and interdisciplinary research.

7.3.1   INDIVIDUAL BENEFITS OF DIALOGUE

In the field of education, there has been much research on the potential for collaborative activities such as dialogue to improve the ability of students to learn new content by integrating new information into their preexisting understandings. This includes research on dialogue as a vehicle for argumentation (e.g., work on dialogical argumentation in Kuhn and Crowell 2011) and research on dialogue from the perspective of human association (e.g., work on "bridging differences" in Nagda 2006). However, rigorous research on the mechanisms and outcomes of dialogue in classrooms is much scarcer than studies describing the types of engagement that have been used (Howe and Abedin 2013; Kuhn 2015; Asterhan and Schwarz 2016). Nevertheless, it is possible to draw some general conclusions about these mechanisms and outcomes.

First, dialogue generally does little to help rote learning of factual content material (Kuhn 2015). As Nokes-Malach et al. (2015, p. 648) observe, "If group members can solve the task individually, then they will show little benefit from collaborating." However, benefits are evident when it comes to ill-structured or complex problems, particularly when there is no single correct answer (Cohen 1994; Asterhan and

Schwarz 2007). This is, in part, because each group member can bring unique knowledge to bear and the dialogue facilitates integration of this diverse knowledge.

A second conclusion is that individual learning gains tend to arise from small group interactions that allow exploratory talk, particularly when participants engage each other directly to explore their individual thinking processes (Kuhn 2015). Students reason more thoroughly and reach better solutions when they engage in social processes (like dialogue) that require them to reflect on their own understanding of a problem and to explain or justify their thinking and reasoning to others (Asterhan and Schwarz 2016). Collaborative forms of argumentation, such as dialogue, in which students work together, explore positions, and make concessions, generate more learning than adversarial forms of argumentation, such as debate (Nussbaum 2008). As Howe and Abedin (2013, p. 343) observe in their review of research on classroom dialogue, a "striking finding is that when [classroom] dialogue involves exchanges of views, all students typically benefit." Collaborative argumentation leads to better mastery of content and the ability to apply knowledge to solve concrete problems.

Why is collaborative argumentation effective in educational settings, in contrast to other forms of verbal exchange? Nussbaum (2008) posits that it generates internal conflict for the individual, triggered by disagreement, creating discomfort and motivation to resolve the discrepancy (cf. Asterhan and Schwarz 2016). This enhanced motivation can lead to deeper engagement with material. Students may also come to realize that their usual patterns of thinking are ineffective, forcing them to practice and adopt new mental models. They may also explore the different viewpoints among them and reason through the discrepancies, thereby leading to greater learning (Asterhan and Schwarz 2016). Thus, more so than many other pedagogical strategies, collaborative argumentation promotes cognitive elaboration and the development of linkages between existing knowledge and new information or perspectives (Nussbaum 2008).

A third conclusion from educational settings is that students may contribute more to a dialogue among peers than to exchanges in which each responds to questions from the teacher (Howe and Abedin 2013). This is best explained from the association perspective on dialogue as being due to the intrinsically engaging nature of dialogue as compared to other forms of classroom interaction. This has been demonstrated in both the educational literature and the organizational science literature (e.g., Tsoukas 2009; Majchrzak et al. 2012), where dialogue has been shown to be more active and engaging for participants as compared to more passive forms of exchange, such as listening to presentations (Kuhn 2015).

So far, we have discussed the impacts of dialogue on an individual's learning of technical content and material in classroom settings. However, another benefit that arises from dialogue is the cultivation of reflexivity by an individual. Reflexivity is a form of "self-distanciation" (Tsoukas 2009) that involves consciously scrutinizing one's tacit knowledge; it has been defined as "the capacity to go on interrogating one's taken-for-granted universe" (Goodyear and Zenios 2007, p. 353). In contrast with problem-related knowledge, such as the chemical composition of a compound, tacit knowledge comprises the assumptions that frame the problem, the types of knowledge one considers relevant to it, and one's epistemological worldview that shapes approaches to problem solving (Polanyi 1967). Tacit knowledge is generally not articulated in language, and individuals may have greater or lesser degrees of self-awareness about the assumptions and values that constitute their "taken-for-granted" reality.

Engaging in dialogue forces one to articulate tacit knowledge explicitly, justifying it to oneself and exposing it to the scrutiny of others (Gonnerman et al. 2015). Kuhn (2015, p. 49) notes that such scrutiny "is valuable precisely because it is so notoriously difficult to carry out on one's own thinking." A dialogue participant may become aware of perspectives that they have never previously considered and/or recognize that the bases of their position are contestable (Kuhn 2015). Dialogue brings the assumptions, boundaries, and limitations of one's position into sharper focus than typically happens in day-to-day interactions, especially when the dialogue involves argumentation.

A final conclusion related to the individual benefits of dialogue is that thinkers who are practiced at collaborative dialogue and reflexivity reap career-long benefits in the form of "epistemic fluency" that "allows one to recognize, appreciate and understand the subtlety and complexity of a belief system that one has not encountered before" (Goodyear and Zenios 2007, p. 358). Today's social and scientific problems require thinkers who can adapt to new types of interactions and ways of thinking. They must be able to do more than just master the concepts and facts of their separate disciplines. Thinkers must also grasp how other cultures and disciplines bring different worldviews to bear on their joint activities, especially when they are working in diverse teams.

7.3.2 THE SOCIAL AND TEAM BENEFITS OF DIALOGUE

The benefits of dialogue can accrue to each participant in a group and to the group as a whole, regardless of whether the group includes students in a classroom, engaged citizens in a public policy deliberation process, or a collection of scientists working on a research problem (Cheruvelil et al. 2014). Of particular interest to TDI are the benefits to teams, i.e., assemblages of people who interact in pursuing collective goals and that exhibit a high degree of role differentiation (Fiore 2008). Team members are dependent upon one another in pursuing a common task, insofar as they must identify their tasks and address them collaboratively to succeed (Saavedra et al. 1993).

The perspectives of argumentation and association help to explain how dialogue can support team reflection and contribute to knowledge integration. To illustrate from the perspective of argumentation, Asterhan and Schwarz (2007) report experiments in which student teams that were instructed to develop arguments for and against alternative solutions to a problem learned more than teams instructed to work together to reach the best solution. Similarly, research in the workplace has shown that the practice of giving a presentation on one's work to a team may be less beneficial than group discourse that clarifies, explains, and explores the connections needed for knowledge integration (Nussbaum 2008). From the association perspective, dialogue improves the capacity of a team to create a shared identity and develop a shared conception of the problem (Salazar et al. 2012). Capitalizing on diversity through collaborative dialogue can lead to positive research outcomes such as novel or more effective solutions (Cheruvelil et al. 2014).

Previous chapters noted that cross-disciplinary research teams, particularly in the early stages of their development as teams, may experience significant cultural and communication challenges working together. Team members may come from varied, potentially incompatible traditions or "epistemic cultures" (Knorr Cetina 1999). As Tsoukas (2009) notes, each team member initially experiences the situation in terms of categories, distinctions, and assumptions that are embedded in their discipline (Polanyi 1967). This can generate difficulties related to disciplinary use of language (Bracken and Oughton 2006; Donovan et al. 2015) and problems integrating different epistemologies (Lélé and Norgaard 2005; Eigenbrode et al. 2007) or mental models (Pennington 2016). Laursen (2018b) notes that these challenges are especially difficult in dialogues that aim to produce collaborative explanations. Understanding and appreciating different epistemic cultures is "cognitively demanding—it involves learning that can sometimes be very difficult" (Goodyear and Zenios 2007, p. 354), but failure to do so can be detrimental to the development of solutions to complex social problems (Della Chiesa et al. 2009).

A cross-disciplinary team must therefore figure out ways to integrate different views of problems and of the world into a coordinated and coherent perspective on their common project; that is, they must combine their individual contributions into a shared representation (Pennington 2016). Teams often launch into problems with a focus on the end goals of actions and solutions, without taking time for initial reflection on how each member conceives the task or understands the group's resources and constraints. In teams composed of members with quite different disciplinary perspectives, a critical first step toward ultimate success is simply becoming aware of their differences (Hall and O'Rourke 2014). Such early interactions can strongly influence the likelihood that they will use their knowledge resources productively to solve problems (Pennington 2016). Through dialogue, "specialists externalize their 'deep knowledge' in a manner that allows others to understand the boundaries, differences, and dependencies between each other's knowledge" (Majchrzak et al. 2012, p. 952). Team members become aware of the partialness of their collective knowledge; this awareness can increase efficiency in solving problems. Dialogue thereby contributes to the argumentational benefit of effective macrocognition in teams: the "internalized and externalized high-level mental processes employed by teams to create knowledge during complex, one-of-a-kind, collaborative problem solving" (Letsky et al. 2007, p. 7).

Effective dialogue within research teams can generate more creative, innovative solutions (Glăveanu 2010). As Tsoukas (2009, p. 949) writes, "group dialogue is important because it helps circumvent a major problem individuals reasoning alone face: generating *alternative* hypotheses, explanations, and theories." Dialogue can also help collaborators make associations between different ideas (Salazar et al. 2012) and create new distinctions through conceptual expansion and conceptual reframing (Tsoukas 2009; Van Knippenberg et al. 2013).

Reflection and macrocognition also provide associational benefits. Team members who engage in constructive dialogue can relate to one another more effectively and respond more effectively to problems they encounter during their work. Establishing norms for productive dialogue early in a joint project increases trust in groups and reduces the chances of interpersonal conflict; when this dialogue includes thinking about and discussing cultural assumptions, more trust arises (Tsoukas 2009; Chua et al. 2012; Majchrzak et al. 2012). By encouraging deep listening and respectful interaction, dialogue fosters modes of behavior and interaction that aid a team in traversing the inevitable difficulties of teamwork. Team members can jointly bridge differences by appreciating other points of view, being critically self-reflective, and engaging each other in constructive ways (Nagda 2006).

7.3.3 CAVEATS

Dialogue, as we have conceptualized it here, can have significant value for teams. However, a few caveats are in order. Dialogue is not always necessary to achieve positive team outcomes (Majchrzak et al. 2012; Asterhan and Schwarz 2016). Indeed, knowledge integration may not even be necessary for some problems that teams tackle. Another caveat is that the benefits of dialogue do not arise from just any type of conversation (Kuhn 2015; Sprain and Black 2018). Despite good facilitation and well-meaning participants, conversation can sometimes be superficial and fail to help participants embrace a collective goal or commit to joint activity. This can occur because participants are unwilling or unable to engage in challenging dialogue (Nagda 2006). For example, airing differences that could engender divisive arguments can impede the goal of getting teammates to work well with each other (Asterhan and Schwarz 2016). For both argumentative and associational reasons, successful participation in productive dialogue also depends on the interpersonal skills of team members (Cheruvelil et al. 2014) and their competence at communication (Jablin and Sias 2001; Thompson 2009).

A further caveat is that dialogue is not always sufficient to ensure positive team outcomes. Even "good" dialogue with positive intentions and genuine engagement can fail to be productive for reasons related to the core intellectual commitments of the participants. For example, differences in worldviews among participants can be irreconcilable. This is especially true when there are differences in core values, such as whether scientists should engage in advocacy. Where groups have major interpersonal challenges or limited collective competence, it may be better first to promote other types of shared experiences that enable the participants to build the common ground necessary for them to be able to reason together through dialogue (Campolo 2005).

7.4 DIALOGUE: TOOLBOX STYLE

As we detailed in Chapter 2, the Toolbox approach uses structured dialogue to enable diverse groups to communicate and collaborate more efficiently and effectively. Members of a diverse group are likely to have different worldviews, as well as different ways of communicating and enacting them. Collaborators who understand a situation differently will have difficulty coordinating their actions. Without an appreciation for these differences, success will depend on luck, which can endanger future success by encouraging false confidence in the team's reasoning capacity (Campolo and Turner 2002; Campolo 2005). The Toolbox approach is designed to increase a team's odds of success by increasing shared understanding of their projects and their teammates and thereby increasing their collective reasoning capacity.

Structured dialogue is the centerpiece of the Toolbox approach, and philosophical analysis figures importantly in how we structure the dialogue through the prompts. The overriding goal in a Toolbox workshop is to increase the mutual understanding of collaborators and enable them to be more effective in conducting their integrative activities, such as cross-disciplinary research. As such, our approach qualifies as a dialogue method for research integration, alongside those discussed by McDonald et al. (2009) (see also Jordan 2016). A dialogue method is a technique for explicitly organizing or structuring conversation to enable participants to share and integrate their perspectives and knowledge (McDonald et al. 2009). Such methods can provide attentional support, enhance understanding, facilitate decision-making, coordinate action, and improve relationships, attitudes, and feelings (Jordan 2016).

McDonald et al. (2009, pp. 6–7) distinguish between "dialogue methods for understanding a problem broadly" and "dialogue methods for understanding particular aspects of a problem." Although the Toolbox approach can be helpful in establishing a broad understanding, it resembles most closely those methods classified under the latter heading, fitting alongside "strategic assumption surfacing and testing," another dialogue method that supports the integration of worldviews (McDonald et al. 2009, pp. 100–101). However, strategic assumption-surfacing and testing is adversarial, privileges dialectical debate, and aims to generate agreement through modification of the assumptions that are discovered in the dialogue, and so differs from the Toolbox approach.

Three elements distinguish the Toolbox approach as a dialogue method. They are adopted because they increase the likelihood that dialogue will be beneficial to participating individuals and teams. These elements are *participant stance*, *structure*, and *facilitation.*

An appropriate *participant stance* is cultivated in a Toolbox workshop insofar as the instructions guide participants toward what Tsoukas (2009, p. 945) calls "relational engagement," thereby fostering the associational benefits participants gain from dialogue. The ground rules and facilitator encourage participants to take responsibility for the discussion, its outcomes, and each other's psychological well-being. Participants are also encouraged to bring their experiences and ideas to the fore and engage in critical self-reflection. Ideally, dialogue leads participants to see "others willing to be honest and confront their biases," which can create "an openness and vulnerability in confronting self and others" (Nagda 2006, p. 563).

A second feature of Toolbox workshops relates to their *structure*, specifically the use of philosophically designed dialogue prompts. These prompts function as boundary objects and therefore help integrate different ways of thinking about a common project (Star and Griesemer 1989). Such a specially designed structure is needed for both argumentative and associational reasons. Viewed from the perspective of argumentation, dialogue is an "effortful process" that reveals, explores, and integrates propositions and conclusions (Van Knippenberg et al. 2013, p. 185). Groups often need assistance to productively engage in these activities (Cohen 1994), and various structures can supply it. The Toolbox approach provides dialogue prompts to structure conversations. These are akin to what Noroozi (2018) calls "sentence openers," which assist participants in developing questions to ask each other; such prompts are a productive way to elicit differing points of view. However, Toolbox prompts are unique in that they are generated using the techniques of philosophical analysis. They are crafted to provide language that captures many of the epistemological and value-related differences that have been observed in cross-disciplinary teams (Lélé and Norgaard 2005; Eigenbrode et al. 2007). By providing an initial vocabulary and spanning a range of epistemological perspectives, they facilitate the process of making the tacit knowledge of each team member available to all team members.

Viewed from the perspective of association, TDI recognizes that people need to explain their knowledge and expertise in ways that resonate with teammates (Pennington 2016). Linking prompts closely to a group's project at the outset could generate reactivity or discomfort if group members hold different views about their work, and so the Toolbox prompts avoid specific project details. Instead, they are designed to address more abstract, conceptual aspects of the project that should have meaning for all of the collaborators, thereby providing an opening that helps participants externalize their tacit knowledge and commitments. As such, the prompts help protect the psychological safety of speakers who may hold a minority viewpoint about the specific work of the team.

In calling the prompts "boundary objects," we emphasize their role in linking the different orientations of cross-disciplinary collaborators. These prompts support multiple interpretations and reveal connections to disciplinary concerns while also being abstract enough to function as a "common object" for the development of shared understanding across disciplines. Despite their abstract nature, they offer the ability for groups to "tack back-and-forth" between the abstract and the local (Star 2010, pp. 604–605). The invocation of the Rwandan genocide in the excerpt in Table 7.1 illustrates this dynamic, with the participants moving from abstract consideration of experimentation and replication to a very specific example and then back to consideration of science in general. By making space for examples in this way, the Toolbox approach is consistent with educational theory that suggests the use of specific examples generates more learning than abstract discussion (Davis 2000).

These boundary object features enable the argumentative and associational benefits described above. Argumentatively, the flexibility of meaning helps participants articulate their reasoning and evaluate that of their colleagues. This flexibility of meaning is grounded in part in the fact that the prompts are written to include vague and ambiguous terminology, as we noted in Chapter 2. By ensuring that different people will interpret the prompts differently, the Toolbox approach invites collaborators to interrogate each other's interpretations and closely examine alternative perspectives. Associatively, Toolbox workshops make space for participants to tell stories, interject humor, and express emotion around the prompts. These modes of talk encourage development of mutual understanding, since as Nagda (2006, p. 563) writes, "appreciating difference is learning about others, hearing personal stories, and hearing about different points of view in face-to-face encounters; it is an openness to learning about realities different from one's own."

Finally, *facilitation* is a key element of a Toolbox workshop. Our approach to facilitation enables a team to own the dialogue that emerges. Two aspects of our facilitation are worth noting. First, the facilitator starts the dialogue with an opening preamble that highlights the nature and value of the dialogue and co-creation activities. The rationale for this is that dialogue about tacit knowledge, mindsets, and the like does not happen in many teams (Pennington 2016) and participants will not necessarily attach any value to such dialogue (Asterhan and Schwarz 2016). The preamble explains the value of such dialogue and why the team is engaging in this teamwork exercise (Cheruvelil et al. 2014). Second, the TDI approach to facilitation is light-handed. This encourages an atmosphere of openness and respect for participant perspectives that is a critical associational element of productive dialogue (Van Knippenberg et al. 2013; Asterhan and Schwarz 2016). Given that "openness" among participants is a goal (cf. Sprain and Ivancic 2017), the facilitator needs to tolerate basic and tangential questions or issues while still keeping the group on task (Pennington 2016). A light-handed approach strikes a middle ground between respecting participants' desires to shape the trajectory of the dialogue and keeping the participants focused on the topics articulated by the prompts and encouraging them to engage one another directly.

Taken together, the unique *participant stance, structure,* and *facilitation* of Toolbox workshops encourage a collaborative approach to the experience that supports openness (an associational benefit) and critical attention to different core beliefs and values (an argumentative benefit). Since this experience is not typically part of the life of many teams, its artificiality can enable collaborators to hit the "reset button" and approach one another anew. This has the effect of encouraging safe argumentation and explicit but respectful disagreement, which can be productive in dialogue. As Salazar et al. (2012, p. 541) note, "Communication of disagreement is more likely to foster knowledge integration and creation" in a collaborative environment than in a contentious one.

7.5 CONCLUSION

In this chapter, we have explained why dialogue has a privileged position in the Toolbox approach. Following Tsoukas (2009) in adopting a structural definition of "dialogue," we distinguish between a perspective on dialogue that emphasizes argumentation and one that emphasizes association, acknowledging the merits of both. The first of these foregrounds the informational dimension of communication, while the latter highlights the sociopragmatic, relational dimension. Both perspectives on dialogue can help explain different aspects of the gains in mutual understanding that a team makes in a Toolbox workshop. By adopting a unique approach to *participant stance, structure,* and *facilitation,* Toolbox workshops regularly achieve the benefits associated with dialogue, including increased critical self-reflection, enhanced appreciation for different assumptions, more accurate awareness of shared knowledge, a greater ability to leverage epistemic diversity, and ultimately a more fully developed and potentially integrated shared understanding of the group's project (Western Michigan University Evaluation Center 2017; see also Chapters 9 and 10).

7.6 ACKNOWLEDGMENTS

We are grateful to past participants in Toolbox workshops for helping us appreciate the structure and impact of dialogue. We also thank Graham Hubbs and Steven

Orzack for their editing assistance. O'Rourke's work on this chapter was supported by the USDA National Institute of Food and Agriculture, Hatch Project 1016959.

7.7 LITERATURE CITED

Aloni, N. 2013: Empowering dialogues in humanistic education. Educational Philosophy and Theory 45:1067–1081.

Asterhan, C. S. C., and B. B. Schwarz. 2007: The effects of monological and dialogical argumentation on concept learning in evolutionary theory. Journal of Educational Psychology 99:626–639.

———. 2016: Argumentation for learning: Well-trodden paths and unexplored territories. Educational Psychologist 51:164–187.

Bakhtin, M. M. 1981: Discourse in the novel. Pp. 259–422 *in* The Dialogic Imagination. University of Texas Press, Austin.

Bangerter, A., and H. H. Clark. 2003: Navigating joint projects with dialogue. Cognitive Science 27:195–225.

Black, L. W. 2008: Deliberation, storytelling, and dialogic moments. Communication Theory 18:93–116.

Bohm, D. 1996: On Dialogue. Routledge, London.

Bracken, L. J., and E. A. Oughton. 2006: "What do you mean?" The importance of language in developing interdisciplinary research. Transactions of the Institute of British Geographers 31:371–382.

Buber, M. 1996: I and Thou. Simon & Schuster, New York.

Campolo, C. 2005: Treacherous ascents: On seeking common ground for conflict resolution. Informal Logic 25:37–50.

Campolo, C., and D. Turner. 2002: Reasoning together: Temptations, dangers, and cautions. Argumentation 16:3–19.

Cherry, E. C. 1957: On Human Communication; a Review, a Survey, and a Criticism. MIT Press, Cambridge, MA.

Cheruvelil, K. S., P. A. Soranno, K. C. Weathers, P. C. Hanson, S. J. Goring, C. T. Filstrup, and E. K. Read. 2014: Creating and maintaining high-performing collaborative research teams: The importance of diversity and interpersonal skills. Frontiers in Ecology and the Environment 12:31–38.

Chua, R. Y. J., M. W. Morris, and S. Mor. 2012: Collaborating across cultures: Cultural metacognition and affect-based trust in creative collaboration. Organizational Behavior and Human Decision Processes 118:116–131.

Cissna, K. N., and R. Anderson. 2002: Moments of Meeting: Buber, Rogers, and the Potential for Public Dialogue. State University of New York Press, Albany.

Cohen, E. G. 1994: Restructuring the classroom: Conditions for productive small groups. Review of Educational Research 64:1–35.

Crowley, S. J., C. Gonnerman, and M. O'Rourke. 2016: Cross-disciplinary research as a platform for philosophical research. Journal of the American Philosophical Association 2:344–363.

Davis, E. A. 2000: Scaffolding students' knowledge integration: Prompts for reflection in KIE. International Journal of Science Education 22:819–837.

Della Chiesa, B., V. Christoph, and C. Hinton. 2009: How many brains does it take to build a new light: Knowledge management challenges of a transdisciplinary project. Mind, Brain, and Education 3:17–26.

Donovan, S. M., M. O'Rourke, and C. Looney. 2015: Your hypothesis or mine? Terminological and conceptual variation across disciplines. SAGE Open 5:1–13.

Eigenbrode, S. D., M. O'Rourke, J. D. Wulfhorst, D. M. Althoff, C. S. Goldberg, K. Merrill, W. Morse, M. Nielsen-Pincus, J. Stephens, L. Winowiecki, and N. A. Bosque-Pérez. 2007: Employing philosophical dialogue in collaborative science. BioScience 57:55–64.

Fiore, S. M. 2008: Interdisciplinarity as teamwork. Small Group Research 39:251–277.

Floyd, J. J. 2010: Listening: A dialogical perspective. Pp. 127–140 in A. D. Wolvin, ed. Listening and Human Communication in the 21st Century. Wiley-Blackwell, Chichester.

Glăveanu, V. P. 2010: Paradigms in the study of creativity: Introducing the perspective of cultural psychology. New Ideas in Psychology 28:79–93.

Gonnerman, C., M. O'Rourke, S. J. Crowley, and T. E. Hall. 2015: Discovering philosophical assumptions that guide action research: The reflexive toolbox approach. Pp. 673–680 in H. B. Huang and P. Reason, eds. The SAGE Handbook of Action Research, 3rd edition. SAGE, Thousand Oaks, CA.

Goodin, R. E. 2005: Sequencing deliberative moments. Acta Politica 40:182–196.

Goodyear, P., and M. Zenios. 2007: Discussion, collaborative knowledge work and epistemic fluency. British Journal of Educational Studies 55:351–368.

Grice, H. P. 1989: Studies in the Way of Words. Harvard University Press, Cambridge, MA.

Habermas, J. 1984: The Theory of Communicative Action, volume 1: Reason and the Rationalization of Society. Beacon Press, Boston, MA.

———. 1996: Between Facts and Norms: Contributions to a Discourse Theory of Law and Democracy. MIT Press, Cambridge, MA.

Hall, T. E., and M. O'Rourke. 2014: Responding to communication challenges in transdisciplinary sustainability science. Pp. 119–139 in K. Huutoniemi and P. Tapio, eds. Heuristics for Transdisciplinary Sustainability Studies: Solution-Oriented Approaches to Complex Problems. Routledge, Oxford.

Howe, C., and M. Abedin. 2013: Classroom dialogue: A systematic review across four decades of research. Cambridge Journal of Education 43:325–356.

Isaacs, W. 1999: Dialogue and the Art of Thinking Together: A Pioneering Approach to Communicating in Business and in Life. Doubleday, New York.

Jablin, F. M., and P. M. Sias. 2001: Communication competence. Pp. 819–864 in F. M. Jablin and L. L. Putname, eds. The New Handbook of Organizational Communication: Advances in Theory, Research, and Methods. SAGE, Thousand Oaks, CA.

Johannesen, R. L. 1971: The emerging concept of communication as dialogue. The Quarterly Journal of Speech 57:373–382.

———. 2002: Ethics in Human Communication. Waveland Press, Prospect Heights, IL.

Jordan, T. 2016: Deliberative methods for complex issues: A typology of functions that may need scaffolding. Group Facilitation: A Research & Applications Journal 13:50–71.

Keyton, J. 1999: Relational communication in groups. Pp. 192–222 in L. R. Frey, D. Gouran, and M. S. Poole, eds. The Handbook of Group Communication Theory and Research. SAGE, Thousand Oaks, CA.

Knorr Cetina, K. 1999: Epistemic Cultures: How the Sciences Make Knowledge. Harvard University Press, Cambridge, MA.

Kuhn, D. 2015: Thinking together and alone. Educational Researcher 44:46–53.

Kuhn, D., and W. Udell. 2007: Coordinating own and other perspectives in argument. Thinking & Reasoning 13:90–104.

Kuhn, D., and A. Crowell. 2011: Dialogic argumentation as a vehicle for developing young adolescents' thinking. Psychological Science 22:545–552.

Kuhn, D., and W. Moore. 2015: Argumentation as core curriculum. Learning: Research and Practice 1:66–78.

Laursen, B. K. 2018a: What is collaborative, interdisciplinary reasoning? The heart of interdisciplinary team research. Informing Science 21:75–106.

————. 2018b: On the intersection of interdisciplinary studies and argumentation studies: The case of inference to the best explanation. Issues in Interdisciplinary Studies 36:93–125.

Lélé, S., and R. B. Norgaard. 2005: Practicing interdisciplinarity. BioScience 55:967–975.

Letsky, M., N. Warner, S. M. Fiore, M. Rosen, and E. Salas. 2007: Macrocognition in complex team problem solving. Proceedings of the 12th International Command and Control Research and Technology Symposium.

Majchrzak, A., P. H. B. More, and S. Faraj. 2012: Transcending knowledge differences in cross-functional teams. Organization Science 23:951–970.

McDonald, D., P. Deane, and G. Bammer. 2009: Research Integration Using Dialogue Methods. Australian National University Press, Canberra.

Nagda, B. R. A. 2006: Breaking barriers, crossing borders, building bridges: Communication processes in intergroup dialogues. Journal of Social Issues 62:553–576.

Nokes-Malach, T. J., J. E. Richey, and S. Gadgil. 2015: When is it better to learn together? Insights from research on collaborative learning. Educational Psychology Review 27:645–656.

Noroozi, O. 2018: Considering students' epistemic beliefs to facilitate their argumentative discourse and attitudinal change with a digital dialogue game. Innovations in Education and Teaching International 55:357–365.

Nussbaum, E. M. 2008: Collaborative discourse, argumentation, and learning: Preface and literature review. Contemporary Educational Psychology 33:345–359.

Pennington, D. 2016: A conceptual model for knowledge integration in interdisciplinary teams: Orchestrating individual learning and group processes. Journal of Environmental Studies and Sciences 6:300–312.

Pickering, M. J., and S. Garrod. 2004: Toward a mechanistic psychology of dialogue. Behavioral and Brain Sciences 27:169–190.

Polanyi, M. 1967: The Tacit Dimension. Routledge & Kegan Paul, London.

Saavedra, R., P. C. Earley, and L. Van Dyne. 1993: Complex interdependence in task-performing groups. Journal of Applied Psychology 78:61.

Salazar, M. R., T. K. Lant, S. M. Fiore, and E. Salas. 2012: Facilitating innovation in diverse science teams through integrative capacity. Small Group Research 43:527–558.

Sprain, L., and S. Ivancic. 2017: Communicating openness in deliberation. Communication Monographs 84:241–257.

Sprain, L., and L. Black. 2018: Deliberative moments: Understanding deliberation as an interactional accomplishment. Western Journal of Communication 82:336–355.

Star, S. L. 2010: This is not a boundary object: Reflections on the origin of a concept. Science, Technology, & Human Values 35:601–617.

Star, S. L., and J. R. Griesemer. 1989: Institutional ecology, "translations" and boundary objects: Amateurs and professionals in Berkeley's Museum of Vertebrate Zoology, 1907–39. Social Studies of Science 19:387–420.

Thompson, J. L. 2009: Building collective communication competence in interdisciplinary research teams. Journal of Applied Communication Research 37:278–297.

Traxler, M. J. 2012: Introduction to Psycholinguistics: Understanding Language Science. Wiley-Blackwell, Malden, MA.

Tsoukas, H. 2009: A dialogical approach to the creation of new knowledge in organizations. Organization Science 20:941–957.

Tuomela, R. 2007: The Philosophy of Sociality: The Shared Point of View. Oxford University Press, Oxford.

Van Den Bossche, P., W. H. Gijselaers, M. Segers, and P. A. Kirschner. 2006: Social and cognitive factors driving teamwork in collaborative learning environments: Team learning beliefs and behaviors. Small Group Research 37:490–521.

Van Knippenberg, D., W. P. van Ginkel, and A. C. Homan. 2013: Diversity mindsets and the performance of diverse teams. Organizational Behavior and Human Decision Processes 121:183–193.

Walton, D. 1999: The new dialectic: A method of evaluating an argument used for some purpose in a given case. ProtoSociology 13:70–91.

———. 2010: Types of dialogue and burdens of proof. Pp. 13–24 *in* P. Baroni, F. Cerutti, M. Giacomin, and G. R. Simari, eds. Proceedings of the 2010 Conference on Computational Models of Argument: Proceedings of COMMA 2010. IOS Press, Amsterdam.

Western Michigan University Evaluation Center. 2017: MSU Toolbox Dialogue Initiative: 2017 evaluation report. Kalamazoo.

8 Best Practices for Planning and Running a Toolbox Workshop

Marisa A. Rinkus
Stephanie E. Vasko

8.1 INTRODUCTION

The Toolbox Dialogue Initiative (TDI) uses an iterative process to plan and implement workshops designed to support the needs and goals of interdisciplinary groups and teams. (Henceforth we use "groups" to refer to the different sorts of collectives that participate in Toolbox workshops.) There are five main steps in this process: information gathering, instrument development, workshop implementation, final reporting, and follow-up reflection. This chapter provides an overview of each step in the process and describes best practices for planning and running a successful Toolbox workshop.

In its early history (i.e., its "analogue phase" described in Chapter 3), TDI used a top-down approach when planning and delivering workshops. In a typical workshop, TDI would use a standard instrument centered on STEM fields—the "Scientific Research Toolbox Instrument" (see Appendix A)—rather than an instrument that reflected input from the participating group. The relationship between TDI and the participating group ended after the delivery of a report (Figure 8.1). Reports also were not standardized and mostly reflected the individual style of the workshop facilitator. More information on such "Toolbox 1.0" workshops can be found in Looney et al. (2014).

This chapter focuses on TDI's more recent history, i.e., its "digital phase." Here we discuss the current "Toolbox 2.0" method of offering workshops (Figure 8.2). This method places more emphasis on the goals of the group participating in the workshop than was the case during the "Toolbox 1.0" phase. The addition of a feedback loop also provides the opportunity for long-term partnerships and sustained capacity-building.

Before embarking on the workshop development process with a group, several logistical matters must be addressed. One is whether the information collected during the workshop will be used for research purposes, or just internally to inform the participating group. If collecting data intended for research purposes

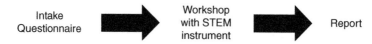

FIGURE 8.1 Toolbox 1.0 workshop development process.

and eventual dissemination through presentations and publications, it will be important to secure human subjects approval through an institutional review board (IRB), independent ethics committee, ethical review board, or research ethics board. TDI has multiple IRB protocols in place at any one time to support its research efforts. Depending on the nature of the institution and the character of the research project, human subjects approval can take a significant amount of time; in light of this, it is best to initiate the process of obtaining it well in advance of the intended workshop.

Another important part of the Toolbox workshop process is securing a venue if you are conducting a workshop in person. Although we acknowledge that sometimes there can be little control over the space where a workshop will occur, the space can be crucial for creating the right environment for dialogue. If multiple dialogue sessions are planned to occur at the same time, it is ideal to use separate rooms. This allows each sub-group to focus on its own dialogue, and it provides better acoustics for good audio recordings, if audio recordings are being made. The space should be large enough to allow the participants to speak to each other face to face. Other considerations include the following: Americans with Disabilities Act (ADA) compliance for facilitators and participants for US-based workshops, Wi-Fi access (if using an online interface), availability of audio-visual equipment for the preamble as well as the presentation and delivery of the co-creation activity, and access to water fountains and restrooms.

FIGURE 8.2 Toolbox 2.0 workshop development process with feedback loop for long-term partnerships.

TABLE 8.1
Example TDI Workshop Goals

Build capacity
Define mission, and/or vision, and/or goals
Identify group and individual values
Identify group and individual priorities
Develop next steps for group action
Identify relevant stakeholders
Reflect on previous process
Foster new intergroup research collaborations
Explore power dynamics within the group

8.2   INFORMATION GATHERING

When preparing a Toolbox workshop, it is best to start gathering information about the partner group at least four to six weeks in advance. This allows time to collect survey responses, clarify participant responses if needed, develop an instrument and request feedback on it, and finalize any logistics for the workshop. We recommend that the bulk of the information used to plan a workshop be solicited through email communications with the group leadership and via a separate web-based survey distributed to expected participants. The survey for the group leadership is intended to disclose more information about the group's history and purpose, why they are interested in a Toolbox workshop, what challenges the group currently faces or perceives that they will face, and the goal(s) for the workshop. Table 8.1 presents a list of goals included in the TDI survey to group leadership. Group leaders are asked to choose up to three goals they would like to achieve during the workshop.

The survey of the leadership is also a good vehicle for obtaining information about workshop logistics, such as the number of participants, the space where the workshop will be held, whether anyone will be participating virtually, or other participant needs (such as ADA compliance). The possibility of virtual participation (discussed at greater length below) is important to address early in the information gathering stage, as it can affect logistics and the dialogue itself.

A survey of the expected participants can occur simultaneously with the leadership survey or after responses to it have been received. The participant survey is designed to identify and better understand *individual* expectations for the workshop, perceived challenges to the group or project, participant roles and relationships, norms of communication, and levels of experience with collaborative or interdisciplinary environments. Participant surveys can be anonymous, thereby allowing individuals to note any areas of conflict and/or issues related, for example, to power dynamics. The survey should consist of no more than ten open-ended questions and serves as the participants' first introduction to TDI and the facilitators.

8.3   INSTRUMENT DEVELOPMENT

TDI has developed numerous Toolbox instruments over the years. The original Toolbox 1.0 instrument is the Scientific Research Toolbox instrument, also known as the STEM (Science, Technology, Engineering, and Mathematics) instrument (see Appendix A). The STEM instrument focuses on fundamental aspects of scientific research knowledge and practice. Rooted in the philosophical underpinnings of scientific research, this instrument is divided into two sections comprising three modules each: the metaphysical section, with modules on reality, values, and reductionism, and the epistemology section, with modules on motivation, methodology, and confirmation (O'Rourke and Crowley 2013). The STEM instrument was primarily used with academic groups and teams (see Looney et al. 2014), but it also had broader application; Schnapp et al. (2012), for example, discuss adapting the STEM instrument for a health sciences context.

TDI currently works closely with groups to design workshop-specific instruments that will best meet their needs. This can involve developing new prompts and modules for an instrument or mixing and matching existing ones from our library of prompts and modules. Instrument development is an iterative process that relies on information and feedback from the group, as well the expertise and experience of TDI facilitators.

The Toolbox instrument is a key element of a Toolbox workshop. It introduces the topics for dialogue and allows for individual reflection and reaction in preparation for the dialogue. A well-developed instrument facilitates the dialogue by stimulating interaction among participants, provoking reflective dialogue, connecting complex themes, and enabling constructive disagreement or agreement. The instrument gets people talking and is designed to help participants avoid responses such as "it depends," "I'm not sure," or "I don't have an opinion about that." (See Chapter 3.) An instrument usually includes three to six modules, with each module containing a core question and six to eight prompts. The core questions are open-ended questions, introducing the module theme and priming participants for the prompts that follow.

The prompts are rating-response statements with response categories of agree to disagree. The statements probe the participants' fundamental beliefs and attitudes and are designed to elicit agreement or disagreement within the group. Toolbox instruments always include "don't know" and "not applicable" response categories. Although these categories enable participants to opt out of expressing agreement or disagreement with an item, they do support the primary purpose of the prompts, which is to stimulate dialogue and reflection within a group. For example, "don't know" allows a participant to signal a range of responses that would not otherwise be available to them, from real ignorance about the issue to concern that the prompt is too vague to assess; these responses can then be used to structure their contributions to the dialogue. Best practices for prompt development based on our experiences are presented in Table 8.2.

8.4 PREPARING FOR THE DAY OF A TOOLBOX WORKSHOP

About five to seven days prior to a workshop, participants should be provided with the workshop agenda and background information on the Toolbox approach. Eigenbrode et al. (2007) and O'Rourke and Crowley (2013) have been supplied, but this book is intended to serve this role in the future. If an online survey is used to collect pre- and post-workshop data, participants should be instructed to bring a device that can connect to the Internet (e.g., tablet, phone, laptop). If hard-copy instruments are used,

TABLE 8.2
Best Practices for TDI Prompt Development

Number of modules	This depends on the length of the dialogue. A 60-minute dialogue could have two to four modules, while a 90-minute dialogue can have up to six modules. These are not strict limits, and it is important to allow sufficient time for all modules to be discussed. If there is a desire to discuss themes in more depth, then fewer modules should be included in the instrument.
Core questions	Each module should have only one or two core questions that articulate the theme of the module. They should be broad, and they should not be logically complex.
Number of prompts	Each module should contain between six and eight prompts. It is important to avoid redundant prompts within a module. Sometimes fewer prompts are better.
Prompt statements	Prompts should present a particular position on an issue so that participants can rate the level of their agreement. Prompts should not admit of "right" answers. Each prompt should only make one claim for the participant to respond to, and so should not be logically complex (e.g., they should not be "and" statements).
Language of prompts	Prompts should be written as positive or negative assertions. Care should be taken in using terms like "could," "would," "can," or "may," which signal types of possibility and could make it too easy to agree or disagree with a prompt. In general, a prompt should not be too easy to agree or disagree with, as this undermines the need to take time to reflect before responding. Terms such as "never," "always," and "must" can be very effective, as they often encourage a strong reaction. It can be helpful to use imprecise language, such as a technical term that has different meanings in different disciplines, as this can lead participants to present their own interpretation or understanding and thereby opens a discussion of multiple perspectives and interpretations.

the appropriate number of copies of the instrument should be made. There should be twice as many copies made as there are participants, and they should be divided into two sets, one labeled "pre" and the other labeled "post." If data are collected from the workshop for research purposes, informed consent forms should be distributed to the participants prior to the workshop, preferably at least a week ahead of time. This allows the participants to fully review the informed consent and present questions or concerns in advance. Participants can provide informed consent the day of the workshop verbally, in writing, or online.

8.5   A TOOLBOX WORKSHOP FROM START TO FINISH

A Toolbox workshop consists of five main segments: preamble, pre/post instrument, dialogue, co-creation activity, and debrief. We discuss each of these in turn and present recommendations for successful implementation. In Table 8.3 we show a

TABLE 8.3
Sample Toolbox Workshop Agenda

Time (min)	Element	Description
10–30	Preamble	Present an introductory overview of the Toolbox approach, the workshop agenda, and the ground rules for the workshop
15	Pre-dialogue instrument	Registers agreement or disagreement with the prompts in the Toolbox instrument for the first time
60–90	Dialogue	Discuss the issues raised by the Toolbox instrument
10–15	Post-dialogue instrument	Registers agreement or disagreement with the prompts in the Toolbox instrument for the second time. Also, allow participants to use the restroom, stretch their legs, etc. after they have completed the post-Toolbox instrument and before starting the co-creation activity
30	Co-creation activity	Conduct an activity that produces an outcome which is valuable for the participants and which is informed by the dialogue
15–20	Debrief	Provide a few thoughts about your observations and open a brief discussion of the process. Example questions include: How was the process? What did you learn about yourself, your discipline, or another discipline?

sample workshop agenda with estimated time allotments for each segment. It is best to schedule three to four hours for a full Toolbox workshop, with a minimum of 60 minutes for the dialogue. As noted above, the longer the instrument is, the longer the dialogue should be.

8.5.1 BEFORE THE DIALOGUE BEGINS: WORKSHOP PREAMBLE, INTRODUCTIONS, AND INFORMATION COLLECTION

The preamble is a short introduction to the Toolbox approach (10 to 30 minutes) that should include a discussion of its conceptual foundations, history, and process. If data are being collected for research purposes, the preamble should also mention the approved IRB protocol and the informed consent statement. The length of the preamble depends on the group's familiarity with the Toolbox approach, interdisciplinarity, team science, dialogue, and the technological readiness of the team if using a web-based interface for the pre- and post-Toolbox instrument. The preamble should also explain the schedule of events for the day, the goal(s) of the workshop, and the workshop ground rules (see Table 8.4). After the preamble has been delivered, the participants should be invited to ask any questions regarding informed consent, the ground rules for the dialogue, or the process in general.

After the preamble, participants are then directed to complete the pre-dialogue instrument. This is the first stage of information gathering during the workshop, so

TABLE 8.4
Sample Discussion Ground Rules for a Toolbox Dialogue

Be kind
Critique ideas, not people
Share your perspective
Speak honestly
Share the airtime—no one dominate
Speak only one person at a time
Choose to be present
Respect other points of view
Keep confidential issues inside the room
Everyone take responsibility for following and upholding the ground rules
End on time

it is important to make sure that enough time is allotted to give the participants an opportunity to reflect on the prompts as they respond to them and to address any possible technological issues that may arise. Before the dialogue begins, we allow participants to introduce themselves to the group if they are not already familiar with one another. This can be especially important if a group is meeting for the first time or has recently added new members. This can also help "break the ice" before the actual dialogue (and the recording) begins.

8.5.2 Workshop Dialogue

The heart of a Toolbox workshop is the dialogue. Once participants have completed the pre-dialogue instrument, the facilitator can open the dialogue by inviting participants to select a prompt, core question, theme, or concept for discussion. There is no right or wrong place to start in the dialogue; participants shouldn't feel pressured to begin with the first prompt or to discuss all of them. The goal is for participants to organically move through the instrument, discussing the aspects that are most salient to them. This may mean spending more time on one topic than another. The facilitator will need to balance the intended purpose of the workshop with the interests and issues that surface in the dialogue. It is important to know at the outset topics that may be of particular interest in case the dialogue needs to be directed to one or a few topics. If possible, the facilitator should not be a member of the group, as this can inhibit the participation of some members. An outside facilitator often works best, as they can moderate the discussion without engaging in the dialogue. The instrument, if well developed, facilitates much of the dialogue, requiring the facilitator to step in only if the discussion has stalled, gotten off track, or if the ground rules have been violated. In Table 8.5, we present our best practices for facilitating a Toolbox workshop.

8.5.3 Workshop Co-Creation

The second part of a Toolbox workshop is a co-creation activity. This refers to an interactive activity that builds on the themes discussed in the dialogue. Common

TABLE 8.5
A Selection of Best Practices for Facilitating a Toolbox Dialogue

Divide workshop participants into dialogue groups of 7 to 12 members each (Krueger and Casey 2000).

Make sure that participants can see each other during the dialogue. This can be accomplished by arranging tables and chairs into a circle, rectangle, or square. You can place tables at an angle so that participants sitting next to each other can make eye contact without leaning in or back.

Have one facilitator, one person to track speaking turns (especially if recording), and one note taker. These should not be people participating in the dialogue. The facilitator can also be the note taker if the group is small and there are no known areas of conflict within the group that would require more facilitation.

The facilitator should be positioned among the participants so that they direct their responses to the other participants. The facilitator should encourage a dialogue that does not include themselves, while remaining present enough to jump in when needed.

Establish ground rules for dialogue either by presenting a predetermined list that can be adapted for the group or by creating them in a collaborative manner with the participants. (See Table 8.4 for a set used by TDI.) Discussion ground rules are a set of behaviors that the group agrees to abide by and uphold for the entirety of the dialogue. These should be on display throughout the dialogue for all participants to see. Several online resources are available to guide creation of ground rules that work for different groups (see Prykucki 2014; University of Kansas Center for Community Health and Development 2020).

Be aware of power dynamics that could require more active moderation during the session. More active moderation may involve interjecting to create space for others to speak, providing alternate ways for participants to register concerns with you (e.g., real-time chat options, survey polls), or having scheduled check-ins every 20 minutes during the dialogue to assess whether the ground rules are being met. If you know power dynamics or hierarchical issues are likely to arise, take steps to ensure that you minimize this in the workshop. Some techniques include the following: have your group remove names tags/badges (if applicable) and try not to mix faculty and graduate students in workshops.

The facilitator should not interject themselves into the dialogue by offering interpretations of the prompts to participants, even if they ask, or by taking sides with a particular participant or a particular response.

Encourage participants to talk to each other, instead of directing comments to the facilitator. In general, dialogues are better when the participants almost forget the facilitator is there and direct the conversation on their own.

goals of such activities include drafting a team glossary, identifying group mission, vision, and goals, or clarifying roles, responsibilities, or next steps within a project. Activities chosen should support the team in making use of insights that arose from the discussion. Sample activities could include those described by Lipmanowicz and McCandless (2014), which are known as *liberating structures* and provide alternative approaches to group work. The overarching goal here is to match your co-creation activity to the goals of your workshop.

8.5.4   Virtual Participation in Toolbox Workshops

Prior to conducting a workshop virtually, there are two main issues to consider: technological readiness and engagement. Drawn from the science of team science, technological readiness concerns the technologies participants are comfortable using and can access (Hall et al. 2008; Olson and Olson 2000). Technological readiness can be assessed via a survey that asks about Internet connection, hardware (e.g., microphones), video capability, and familiarity with platforms such as Zoom® and Skype®. Both the technological readiness of the participants and the technologies selected for the workshop will influence the ways in which participants engage in the workshop.

To ensure participant engagement in a workshop, it is important to establish whether all participants will join the workshop virtually or if there will be a mix of in-person and virtual participants. In our experience, virtual participation by all is better than a mix. If a session is mostly in-person with a few virtual participants, it can be challenging for virtual participants to feel as though they are part of the dialogue. If a mix of virtual and in-person participants is unavoidable, set up the meeting room and cameras so that virtual participants can see all the participants and vice versa (Hampton et al. 2017). External microphones should be used to ensure the quality and projection of the sound. As a facilitator, it is important to check in regularly with virtual participants, either through a chat function or by pausing the discussion.

Technological readiness and engagement come together when selecting a virtual platform. A virtual platform that has audio, video, and chat functions offers multimodal options that can accommodate diverse participant needs. TDI believes that it is important to have the video function so that participants can see each other, which can help the dialogue flow naturally. However, a Toolbox workshop can be delivered via conference call if necessary. It is essential to be familiar with the various options in advance so that you can troubleshoot if needed. If you plan to record the session you must obtain consent from every participant. We urge you to make your workshop accessible by having live-captioned video, making handouts available before the workshop, and choosing platforms that support the use of a screen reader.

When considering technological readiness and engagement in your workshop design, it is important to consider the impact of the workshop on the participants. A typical Toolbox workshop lasts three to four hours. This can be taxing, especially in a virtual environment. There are several options for adapting the workshop design to a virtual environment: reducing the length of the workshop by providing a preamble in advance, reducing the length of the dialogue, and conducting the co-creation activity in another meeting. Be sure to remind participants that they can take a break or turn off their cameras as needed. Remember to start and end on time, but allow time to address technological issues that may arise. Given the ever-changing nature of technology, careful consideration of technological readiness and engagement provides a good starting point for the design of a successful virtual workshop.

8.5.5   Feedback Questionnaire

Within two weeks after completion of the workshop, a feedback questionnaire is distributed to collect information regarding participant reactions, learning, and post-workshop behaviors. The responses help determine whether the group or team has made progress toward their workshop goal(s) and also inform recommendations to be included in the final report. It also allows for participant evaluation of the instrument, dialogue, co-creation activity, and facilitation.

8.6   REPORT AND REFLECTION

After a workshop, the information collected by the TDI members is then synthesized into a TDI report, which typically contains a workshop summary, highlights from the dialogue session, pre- and post-dialogue instrument scores, description and results of the co-creation activity, and recommendations. The workshop summary presents the context for the workshop and dialogue session(s) and a general description without identifiers of who was involved. The highlights of the dialogue session include recurrent themes from the dialogue, based on facilitator notes and the audio recording. More detailed comments from participants regarding the prompts and themes can be included in an appendix. The section on pre- and post-dialogue instrument scores presents general findings from the responses to the Toolbox prompts, highlighting shifts in responses such as from "don't know" in the pre-dialogue instrument to "agree" or "disagree" in the post-dialogue instrument. It may also be important to note where responses were divided between "agree" and "disagree" even after the dialogue, as these may be areas that require further discussion. Full summary of frequency counts for each set of pre and post prompt responses by module can also be included in an appendix. It is important to note in the report that because these prompts are meant to spur discussion and are not psychometrically validated measures, the interpretation here cannot be generalized beyond the participants of this workshop.

A brief description of the co-creation activity follows, outlining how the co-creation activity built on the Toolbox dialogue and what it involved. The information generated from the co-creation activity is collated and presented in an appendix for reference. Results from the feedback questionnaire can also be included in an appendix. Finally, the report concludes with a list of recommendations for consideration based on the information collected before, during, and after the workshop. These recommendations are consistent with the goal of the workshop and present suggestions of resources, when applicable. Reports are written and sent to participants no later than two months after the workshop. Once the final report is distributed to participants, a follow-up meeting with the group leader or whole team can be scheduled to discuss the report and plan for future workshops. The reports serve as a resource for groups to reflect back on their dialogue, continue conversations, or start new ones.

8.7   CONCLUSION

This chapter outlines the steps in planning and implementing a Toolbox workshop and our recommendations for best practices. It is important to remember that every workshop has its own unique aspects that should be taken into consideration. Our goal in providing best practices is to draw attention to the key elements of a Toolbox workshop and provide guidance based on our experience. Toolbox workshops can be a journey of exploration for the participants and the roles of the facilitator are to create an inclusive and engaging workshop and to guide participants along the way.

8.8   ACKNOWLEDGMENTS

The authors would like to acknowledge the many people that have been involved in organizing and running Toolbox workshops over the years as well as members of the TDI community that have helped to usher in its new era.

8.9   LITERATURE CITED

Eigenbrode, S. D., M. O'Rourke, J. D. Wulfhorst, D. M. Althoff, C. S. Goldberg, K. Merrill, W. Morse, M. Nielsen-Pincus, J. Stephens, L. Winowiecki, and N. A. Bosque-Pérez. 2007: Employing philosophical dialogue in collaborative science. BioScience 57:55–64.

Hall, K. L., D. Stokols, R. P. Moser, B. K.Taylor, M. D. Thornquist, L. C. Nebeling, C. C. Ehret, M. J. Barnett, A. McTiernan, N. A. Berger, M. I. Goran, and R. W. Jeffery. 2008: The collaboration readiness of transdisciplinary research teams and centers: Findings from the National Cancer Institute's TREC year-one evaluation study. American Journal of Preventive Medicine 35(2 Suppl): S161–S172.

Hampton, S. E., B. S. Halpern, M. Winter, J. K. Balch, J. N. Parker, J. S. Baron, M. Palmer, M. P. Schildhauer, P. Bishop, T. R. Meagher, and A. Specht. 2017: Best practices for virtual participation in meetings: Experiences from synthesis centers. The Bulletin of the Ecological Society of America 98(1), 57–63.

Krueger, R. A., and M. A. Casey. 2000: Focus Groups: A Practical Guide for Applied Research. SAGE, Thousand Oaks, CA.

Lipmanowicz, H., and K. McCandless. 2014: The Surprising Power of Liberating Structures: Simple Rules to Unleash a Culture of Innovation. Liberating Structures Press, Seattle.

Looney, C., S. Donovan, M. O'Rourke, S. J. Crowley, S. D. Eigenbrode, L. Rotschy, N. A. Bosque-Pérez, and J. D. Wulfhorst. 2014: Seeing through the eyes of collaborators: Using Toolbox workshops to enhance cross-disciplinary communication. Pp. 220–243 in M. O'Rourke, S. J. Crowley, S. D. Eigenbrode, and J. D. Wulfhorst, eds. Enhancing Communication and Collaboration in Interdisciplinary Research. SAGE, Thousand Oaks, CA.

Olson, G. M., and J. S. Olson. 2000: Distance matters. Human-Computer Interaction 15(2/3):139–178.

O'Rourke, M., and S. J. Crowley. 2013: Philosophical intervention and cross-disciplinary science: The story of the Toolbox Project. Synthese 190:1937–1954.

Prykucki, B. 2014: *From ground rules to shared expectations*. Downloaded from www.canr. msu.edu/news/from_ground_rules_to_shared_expectations.

Schnapp, L. M., L. Rotschy, T. E. Hall, S. J. Crowley, and M. O'Rourke. 2012: How to talk to strangers: Facilitating knowledge sharing within translational health teams with the Toolbox dialogue method. Translational Behavioral Medicine 2:469–479.

University of Kansas Center for Community Health and Development. 2020: *Group Facilitation and Problem Solving*. Downloaded from https://ctb.ku.edu/en/table-of-contents/leadership/group-facilitation.

9 Enhancing Cross-Disciplinary Science through Philosophical Dialogue

Evidence of Improved Group Metacognition for Effective Collaboration

Brian Robinson
Chad Gonnerman

9.1 INTRODUCTION

The Toolbox method for enhancing team effectiveness has been used in over 300 workshops since 2005. Do Toolbox workshops improve teams' collaborative capacity? This chapter begins to answer that question by providing a quantitative analysis of data from Toolbox workshops. (The next chapter will continue this effort with a qualitative analysis.) It is important to specify what *improving collaborative capacity* means and how we measure success. By examining participants' responses to the same prompts before and after their dialogue, we can look for (1) belief revision and (2) belief formation. In a Toolbox workshop, two different categories of beliefs are at play (see Figure 9.1). Each participant is asked to think about their own research-relevant worldview and about the relationships among the diversity of research-relevant worldviews in the group (i.e., *group metacognition*, which we describe in greater detail below). We follow O'Rourke and Crowley (2013, p. 1938) in understanding "worldviews to be sets of more or less tacit beliefs held by researchers about what they are studying and how to study it, as well as views about the nature of the output of their inquiry." As they point out, Toolbox workshops aim to encourage examination of and dialogue about participants' worldviews. Different prompts inquire about different research-relevant beliefs in participants' worldviews. So, we will look for belief revision and belief formation within a workshop at the level of the individual and at the level of the team. Finally, we will analyze participants' own feedback on the workshop as a further means of assessing belief revision and belief formation.

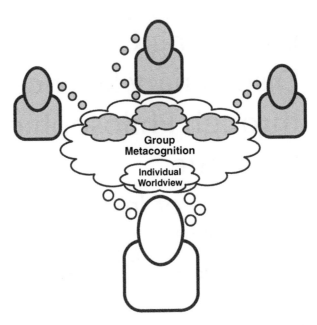

FIGURE 9.1 In a Toolbox workshop, each participant has their own beliefs about specific prompts and beliefs about everyone else's beliefs (also called "group metacognition").

9.2 DATA

Each Toolbox dialogue begins and ends with participants rating the strength of their agreement or disagreement with a series of prompts that are associated with a five-point rating response scale (1 = Disagree, 5 = Agree). These prompts are organized into modules, with each module's theme indicated by a core question. In between, participants engage in a facilitated dialogue based on these prompts. Comparing rating responses from before and after the dialogue allows for assessment of the dialogue's effects on the participants' beliefs.

The Toolbox instrument has varied over the more than 300 workshops conducted to date. For present purposes, we are limiting our analysis only to those groups that used the Scientific Research Toolbox instrument, or "STEM" (Science, Technology, Engineering, and Mathematics) Toolbox instrument, which is the oldest and most frequently used (Looney et al. 2014). The modules (and corresponding core questions) in the STEM Toolbox instrument are as follows: Motivation (*Does the principal value of research stem from its applicability in solving problems?*), Methodology (*What methods do you employ in your disciplinary research?*), Confirmation (*What types of evidentiary support are required for knowledge?*), Reality (*Do the products of scientific research more closely reflect the nature of the world or the researchers' perspective?*), Values (*Do values negatively impact scientific research?*), and Reductionism (*Can the world under investigation be reduced to independent elements of study?*). See Appendix A for the full text of every prompt in the STEM Toolbox instrument.

Each module has four or five prompts that develop the core question for that module. We refer to these as *individual-views prompts*, since they concern individual participants' research-relevant beliefs. The STEM Toolbox instrument has 28 individual-views prompts across all six modules. Each module has a final prompt that asks, "Do the members of this team have similar views concerning the [module name]'s core question?" We refer to these six prompts as the *similar-views prompts*, since they are designed to assess each participant's beliefs about the similarity between their own views and those of the other team members on the module core questions. This sort of thinking about what the others in the group are thinking is *group metacognition*. We will be considering the individual-views and the similar-views prompts separately.

Some of the workshops have been conducted with pre-existing research groups or teams, while others have been conducted with ad hoc groups of interested participants (including academics and non-academics). To evaluate the effectiveness of the Toolbox method for enhancing communication and collaboration within research teams, we analyze only the data from the 20 workshops with pre-existing research groups that used the STEM Toolbox instrument. The mean group size was approximately seven participants; there were 141 total participants across all 20 workshops. (The de-identified dataset is available at: https://osf.io/epx2u/.)

9.3 INDIVIDUAL WORLDVIEWS

9.3.1 BELIEF REVISION

One way that the Toolbox method can be effective is if it encourages participants to revise their philosophical worldviews during a dialogue with colleagues who support and argue for other worldviews. We can test whether such a shift has occurred by comparing the mean responses to any given prompt from before and after the dialogue. We conducted paired t-tests to compare the means for each prompt from before and after the dialogue. The results are shown in Table 9.1. Since each of the 28 prompts is tested separately, the probability of a false positive (i.e., the family-wise error rate) is quite high. Using the Holm-Bonferroni method to control for family-wise error (Holm 1979), the pre/post-dialogue difference in means was significant only for the prompts Methodology 1 and Confirmation 3. Even here, the changes in mean five-point rating responses from before to after the dialogue were small, $d = -0.24$ and -0.31 respectively. For all other prompts, the differences were even smaller.

This result, however, is not surprising. Academic researchers (including those participating in Toolbox workshops) have well-established worldviews on the nature of their research, which often have been molded and reinforced through their disciplinary training as undergraduates, graduate students, and early career researchers (or longer) (Eigenbrode et al. 2007; O'Rourke and Crowley 2013). These worldviews are unlikely to change much in a two-hour dialogue.

Moreover, the absence of individual belief change is not evidence that Toolbox workshops are ineffective. The primary objective of a Toolbox workshop is not

TABLE 9.1

Mean Five-Point Rating-Response Scores for Individual-Views Prompts*

Prompt	Pre M	Pre SD	Post M	Post SD	t	df	p	d
Motivation 1	3.76	0.98	3.59	1.04	2.42	147	.017	−0.17
Motivation 2	2.99	1.24	2.97	1.23	0.18	145	.859	−0.02
Motivation 3	2.71	1.29	2.75	1.31	−0.60	140	.550	0.03
Motivation 4	3.94	1.10	3.80	1.11	1.67	145	.098	−0.13
Methodology 1	3.26	1.32	2.96	1.22	3.65	148	<.001	−0.24
Methodology 2	3.88	1.23	3.88	1.23	0.00	146	1.000	0.00
Methodology 3	2.29	1.17	2.42	1.15	−1.97	144	.051	0.11
Methodology 4	3.31	1.32	3.24	1.36	0.99	144	.324	−0.05
Methodology 5	2.99	1.16	2.97	1.14	0.30	145	.764	−0.02
Confirmation 1	3.82	0.97	3.59	1.05	2.98	146	.003	−0.23
Confirmation 2	3.75	1.00	3.62	1.00	1.52	141	.131	−0.13
Confirmation 3	3.65	1.27	3.26	1.28	4.13	146	<.001	−0.31
Confirmation 4	3.87	0.93	3.90	0.92	−0.44	143	.664	0.03
Confirmation 5	4.44	0.86	4.41	0.75	0.43	145	.671	−0.04
Reality 1	3.67	1.18	3.61	1.13	0.67	146	.505	−0.05
Reality 2	2.95	1.26	3.01	1.15	−0.60	133	.553	0.05
Reality 3	3.19	1.21	3.36	1.13	−2.08	138	.039	0.15
Reality 4	3.18	1.42	3.32	1.36	−1.73	132	.087	0.10
Values 1	1.94	1.14	2.04	1.18	−0.86	141	.392	0.09
Values 2	1.98	0.92	2.03	0.88	−0.63	146	.533	0.06
Values 3	2.81	1.32	2.62	1.21	1.99	143	.048	−0.15
Values 4	3.35	1.26	3.35	1.15	0.00	138	1.000	0.00
Values 5	2.68	1.23	2.60	1.03	0.78	129	.440	−0.07
Reductionism 1	2.89	1.22	2.79	1.15	0.95	119	.345	−0.08
Reductionism 2	2.05	0.99	2.18	1.11	−1.70	135	.091	0.12
Reductionism 3	3.68	0.97	3.71	0.98	−0.38	140	.737	0.03
Reductionism 4	2.98	1.35	3.22	1.11	−2.50	141	.014	0.19
Reductionism 5	4.48	0.69	4.32	0.71	2.54	144	.012	−0.23

* Mean five-point rating response scores and standard deviations for each individual-views prompt from both before and after the dialogue, as well as the results of paired *t*-tests comparing the pre- and post-dialogue responses for each prompt. Using the Holm-Bonferroni method to control for family-wise error, the pre/post-dialogue difference in means was significant only for Methodology 1 and Confirmation 3. *df* denotes degrees of freedom, *p* denotes the probability of the observed test statistic, and *d* denotes the effect size.

individual belief revision (Eigenbrode et al. 2007; O'Rourke and Crowley 2013). Rather, the aim is to foster integration in a team by enhancing collaborative capacity. For more on the nature of and central role of integration in team science and inter- and transdisciplinary research, see Bammer (2013), Brigandt (2010, 2013), Gerson (2013), Klein (1990, 2012), and Repko (2007), along with Chapter 5. Integration does not require individual belief revision (O'Rourke et al. 2016). You and I may have several strongly conflicting beliefs related to our research but still be able to effectively collaborate with one another so long as we (a) are aware of our differences and (b) develop methods for managing and respecting these differences in a productive way. This may require one or both of us to revise our beliefs about the other's beliefs. (This kind of belief revision is not, however, a kind of individual belief revision, since neither of us modifies our worldviews in that case, only our beliefs about the other's worldview.) Consequently, the measure of the effectiveness of the Toolbox method does not lie in its ability to foster individual belief revision. We must look elsewhere.

9.3.2 Belief Formation

When participants have a pre-existing worldview, those beliefs tend not to shift much during the dialogue. Sometimes, however, a participant does not come into a Toolbox workshop with a long-established belief on a relevant philosophical issue related to their research. For example, consider the prompt Reality 2 ("Scientific claims need not represent objective reality to be useful"). While a scientist might already have set views on scientific realism generally, they might not have considered this particular question before. In that case, they might respond, "I don't know" to this prompt rather than by rating agreement or disagreement with it on the five-point rating scale.

One place, then, that we can look for the influence of the Toolbox method is on individual belief *formation*. To examine individual belief formation we compared the number of participants that responded, "I don't know" to each of these prompts before the dialogue to how often they responded that way after the dialogue. (See Figure 9.2.) Bars with a positive value in Figure 9.2 show belief formation for that prompt; fewer participants responded "I don't know" after the dialogue. For

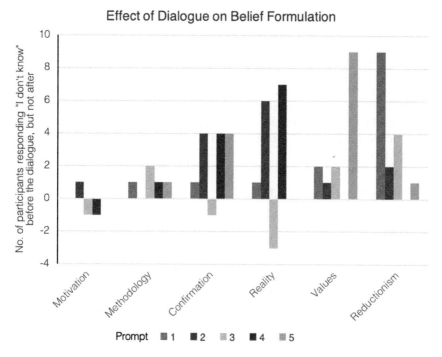

FIGURE 9.2 Shifts in the number of participants responding "I don't know" to each individual-views prompt before and after the dialogue, grouped by module.

instance, for the prompt Values 5 ("Allowing values to influence scientific research is advocacy"), five fewer participants responded with "I don't know" after the dialogue than did beforehand. For all prompts combined, a paired t-test indicates that the number of "I don't know" responses after the dialogue was significantly less than before ($t(27) = 3.68$, $p < 0.001$). (It is important to note that this test only determined that there was statistically significant fewer "I don't know" responses after the dialogue than before across the entire Toolbox STEM instrument, not a test of individual prompts. Such prompt-level testing would require a much larger sample size, and would also be superfluous to demonstrating that the Toolbox method is an effective means of belief formation in general.) This result suggests that Toolbox workshops are an effective method of fostering individual belief formation.

One further point is warranted. Figure 9.2 depicts the shifts in the number of participants responding, "I don't know" to each individual-views prompt before and after the dialogue. Four prompts (Motivation 3, Motivation 4, Confirmation 3, and Reality 3) have negative values, indicating that *more* participants responded, "I don't know" after the dialogue than beforehand. These results suggest that the dialogue encouraged belief destruction rather than belief formation for a *few* participants. While belief formation is certainly indicative of a successful Toolbox workshop, an appreciation for the purpose and value of philosophy shows that belief destruction can also be a mark of success. In Plato's *Apology*, Socrates argues that his wisdom amounts to knowing that he knows nothing (as opposed to thinking that he knows something when he does not). In most of Plato's dialogues, Socrates continues to question his conversational partner until that person is brought to a state of aporia, in which the partner's initial view is obviously self-contradictory and untenable. Russell (1997) claims that the value of philosophy resides in the uncertainty it provides, i.e., it combats dogmatism by prompting us to question beliefs we've previously left unexamined. When a participant changes from initially having a view on a prompt (by responding 1–5 to rate disagreement or agreement with it) but afterward professes ignorance, it may indicate a philosophical success, a recognition that one's initial belief perhaps wasn't as well supported or well-conceived as originally thought.

9.4   GROUP METACOGNITION

9.4.1   BELIEF FORMATION

As noteworthy as these findings on individual belief formation are, much more important for the evaluation of the Toolbox method is the effect it has on group metacognition. Metacognition is thinking about thinking (Flavell 1979; Metcalfe and Shimamura 1994). If you have a thought and then consider why you have that thought, you are engaged in a metacognitive activity. Asking yourself if what you think you remember is really what happened is also metacognition, as is wondering about the nature, certainty, or sources of things you think you know. The nature, function, basis, and applications of metacognition are an increasingly important topic in a variety of sub-fields in psychology. Several researchers of late in educational psychology, for instance, have focused on the role of metacognition in language learning by K1/n12 or ESL/EFL (English-as-Second/Foreign Language) students (Boulware-Gooden et al. 2007; Winne 2017).

Typically, though not necessarily always, metacognition is discussed in terms of thinking about *one's own* thinking. Responding to the individual-views prompts requires this kind of metacognition. Another kind of metacognition is thinking about *others'* thinking, i.e., group metacognition (Chalmers 2009; Siegel 2012; Biasutti and Frate 2018; Teng 2020; Teng and Reynolds 2019). An individual exhibits group metacognition when they think about what others in the group think. As we noted above, a central goal of Toolbox workshops is to enhance a group's mutual understanding, which is essential for integration (see Chapters 2, 5, and 10). This sort of mutual understanding requires individuals to think about what other group members think (i.e., their worldviews) and how those relate to their own worldview, i.e., group metacognition.

Our method for assessing group metacognition involves the use of the similar-views prompts at the end of each module. For these prompts, participants rate their agreement or disagreement (1 = Disagree, 5 = Agree) with the claim that everyone in the group has a similar view on the module's core question. We also offer participants the options of "I don't know" and "N/A" here (as on all prompts). The core questions (and the individual-views prompts) address philosophical topics that team members may never have discussed explicitly before. So, it is plausible that on many topics, participants will not initially know how similar team members' philosophical worldviews are prior to their Toolbox workshop. Consequently, we begin by examining belief formation in group metacognition rather than belief revision. Belief formation in group metacognition occurs when an individual initially has no belief about the similarities of their views to those of their group for a module, but comes to have a belief about the similarity of their beliefs compared to everyone else. Evidence of belief formation in group metacognition after a group's dialogue is indicative of enhanced mutual understanding.

As we did when investigating individual belief formation, we compared the number of participants who responded, "I don't know" before and after the dialogue, only this time focusing on each of the six similar-views prompts. The results (see Figure 9.3) shows a marked difference after the dialogue. Of the 141 participants, at least 18 fewer (12.8%) responded, "I don't know" to each similar-views prompt after the dialogue. For the Values module, 41 fewer participants (29%) responded in this way after as compared to before the dialogue. The results suggest that these research teams did *not* have a mutual understanding of one another's worldview prior to the workshop, particularly about the role values should play in scientific research.

9.4.2 BELIEF REVISION: IMPROVED GROUP METACOGNITION

To examine the effectiveness of Toolbox workshops at enhancing mutual understanding, it is not enough to show an improvement in belief formation in group

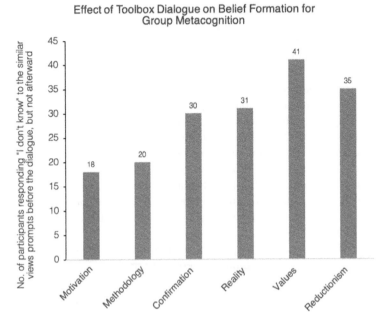

FIGURE 9.3 Shifts in the number of participants responding "I don't know" to each similar-views prompt before and after the dialogue, grouped by module.

metacognition. While removing ignorance about the beliefs of others in a group is a critical step in enhancing mutual understanding, it also matters how accurate those group metacognitive beliefs are. Assessing the accuracy of group metacognition requires three measurements. First, for each team, we examined how similar all their responses were for the individual-views prompts in each module. This tells us how similar their views *actually* are. Second, we examined how similar they *think* their views are. Third, accuracy is measured based on how closely the group's assessment of their similarity corresponds to their actual similarity. This analysis was performed for each team and for each module for both pre- and post-dialogue data. If, in general, the accuracy of group metacognition improves after the dialogue, then we take that as evidence of the effectiveness of Toolbox workshops at enhancing mutual understanding.

How we performed this analysis warrants elaboration. Since we separately analyzed the shifts in the "I don't know" responses to test for belief formation, we removed all of these responses (as well as any "N/A" responses, which were less than 1%). Then we separated the pre-dialogue responses from post-dialogue responses. To determine how similar a group thinks their views are for a module, we examined their responses to the similar-views prompts for that module. The similar-views prompts ask how similar a team's views are on a module's core question. We calculated the mean five-point rating response for each similar-views prompt (one per module) for each team, both before and after the dialogue.

The core questions, however, are open ended without quantitative responses. We do not gather data *directly* about participants' views on the core questions, nor should we. They are multi-faceted by design, addressing a range of related philosophical questions. The individual views prompts in each module are designed to develop different facets of the core question. Consequently, assessing the similarity of a group's views on the core question requires constructing a composite from their responses to the individual-views prompts. To that end, we calculated the standard deviation for each prompt for each team in order to measure how similar team members' views were for that prompt. We then calculated the mean standard deviation for each module for each team. This method provides the measurement of actual similarity of a group's individual beliefs for that module (either before or after their dialogue).

For the pre-dialogue responses, each team has six mean standard deviation measurements, one per module; the same is true for the post-dialogue data. Likewise, each team has a mean pre-dialogue rating response for each of the six similar-views prompts and six more for the post-dialogue responses. We then paired these scores for each module and for each team (mean standard deviation for individual-views prompts and mean response for the corresponding similar-views prompt). In other words, a team has six pairs, one per module, for their pre-dialogue responses. They have another six post-dialogue pairs. With 6 modules and 20 teams, that produces 120 pairs for the pre-workshop data and 120 pairs for the post-workshop data. (See Figure 9.4.)

As we noted, there were high rates of "I don't know" responses to the similar-views prompts, especially in the pre-dialogue data. Too few participants on a team rating their agreement with the similar-views prompts would be an insufficient basis to draw any conclusions. A few teams had fewer than four participants respond 1–5 (i.e., rate their agreement with the statement in the five-point rating scale). When this occurred,

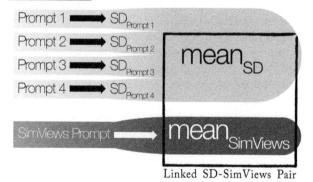

FIGURE 9.4 The association between the mean standard deviation (mean$_{SD}$) for responses to the individual-views prompts and the team's mean five-point rating response to the module's similar-views (SimViews) prompt. For each team, six pre-dialogue linked pairs were calculated and six post-dialogue pairs.

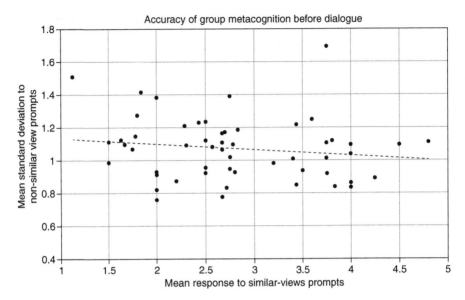

FIGURE 9.5 The correlation between the similarity of a team's responses to a module's individual-views prompts (measured by mean standard deviation) and a team's mean response to the similar-views prompt for the same module before the dialogue.

the linked standard deviation-similar views pair involving that module was removed from consideration. This left 55 pre-dialogue pairs and 103 post-dialogue pairs.

To assess the accuracy of participants' responses to the similar-views prompts, we calculated Pearson product-moment correlation coefficients between the pairs. Prior to the dialogue, there was no significant correlation ($\alpha = 0.05$) between the similarity of a team's view on a module and their assessment of the similarity of their views ($r(55) = -0.16$, $p = 0.24$). (See Figure 9.5.) After the dialogue, there was a significant correlation ($[r(103) = -0.22$, $p = 0.03$). (See Figure 9.6.) The R-squared for this correlation (0.05) implies that 5% of the variance in dialogue participants' assessments of the similarity of their views is accounted for by the similarity of the team's views on matters probed by the individual prompts constituting the rest of the modules. The post-dialogue result for the accuracy of group metacognition has a small effect size (Cohen 1988).

We can draw two important conclusions from these results. First consider only the post-dialogue result. This result suggests that after the dialogue, teams were moderately accurate in assessing the similarity or dissimilarity of philosophical worldviews among everyone in the group. To contextualize this conclusion, consider the following hypothetical. Suppose, for example, that for a given module, the mean standard deviation for a team's responses to the individual-views prompts was nearly 2.0, its theoretical maximum on a five-point rating response scale. This would imply that the team is very divided on the prompts in this module. There would be an almost even split of participants responding 1 (Disagree) as responding 5 (Agree) to each prompt. If the team accurately recognizes this difference, then they would

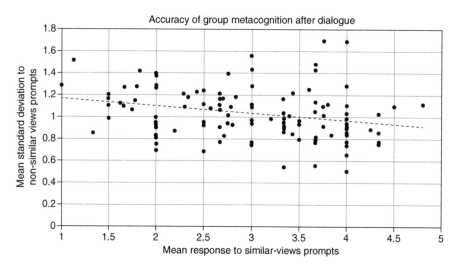

FIGURE 9.6 The correlation between the similarity of a team's responses to a module's individual-views prompts (measured by mean standard deviation) and a team's mean response to the similar-views prompt for the same module after the dialogue.

respond to the similar-views prompt with 1 (Disagree), indicating they disagree with the prompt's claim that the team has similar views. Alternatively, a team with a mean standard deviation of 0.0 for their responses to the individual-views prompts would completely agree about that module. They would accurately assess their agreement by everyone responding 5 (Agree) to the similar-views prompt for that module. In other words, accurate assessments of agreement entail an inverse relationship between mean standard deviation and mean responses to the similar views prompt (hence the negative r value). As mean standard deviation goes up, accurate responses to the similar views prompt go down, and vice versa. The estimated correlation ($r(103) = -0.22, p = 0.03$) is consistent with this inverse relationship. Cohen (1988) classifies any result whose absolute value is greater than 0.1 but less than 0.3 as small. The effect of a Toolbox workshop appears significant but not large. This is to be expected. Over a two-hour Toolbox dialogue, teams do not and cannot cover everything. It would be surprising if only a two-hour dialogue could effect a large improvement in group metacognition by itself. The Toolbox method is not a panacea, but this result suggests that it enhances mutual understanding.

The second conclusion we can draw concerns whether teams had enhanced mutual understanding after their dialogues or rather they already were fairly accurate in their group metacognition before their Toolbox workshops. To answer this question, we compared the Pearson's correlations from before and after the dialogues. As Figure 9.6 shows, after the dialogue the correlation is more negative, indicating that teams with decreasingly similar views on a module had increasing rates of disagreement with the similar views prompt. In short, participants are more accurately assessing the similarity of their teammates' worldviews after the dialogues. This is the result we would hope to see if Toolbox workshops enhance mutual understanding.

9.5   POST-WORKSHOP PARTICIPANT EVALUATIONS

We also evaluated the success of the Toolbox method by examining participants' post-workshop feedback. A few weeks after a Toolbox workshop, participants are emailed a follow-up post-workshop questionnaire to assess their views on the workshop experience. The format of the questionnaire has evolved over time, so we cannot use the same dataset that we analyzed above. The current format was used with six research teams that also used the STEM Toolbox instrument. Each participant ($n = 74$) was asked to rate their agreement with the prompts listed in Table 9.2 on a five-point rating response scale (1 = Strongly Disagree; 5 = Strongly Agree). The results are summarized in Figure 9.7.

TABLE 9.2
Prompts Used in the Post-Workshop Questionnaire

Prompt Number	Question
Q1	Overall, I enjoyed the Toolbox workshop.
Q2	The Toolbox workshop helped me clarify some of my research worldviews.
Q3	The Toolbox prompts were effective conversation starters.
Q4	Our conversation was an open exchange of thoughts and ideas.
Q5	I felt free to present a view that differed from others in my group.
Q6	My group's discussion addressed issues that I personally wanted to discuss.
Q7	The Toolbox prompts stimulated new thoughts for me about differing research worldviews.
Q8	Comments from others stimulated new thoughts for me about differing research worldviews.
Q9	I have a better understanding of other group members' research because of the Toolbox dialogue.
Q10	The Toolbox workshop helped our group identify some of our differing research worldviews.
Q11	The Toolbox workshop helped build a sense of community among the group.
Q12	After the workshop, I have a better idea about how to communicate with researchers with differing research worldviews.
Q13	Based on my experience in the Toolbox workshop, I think mutual understanding of differing research worldviews is important for productive collaboration in cross-disciplinary teams.
Q14	Participating in the Toolbox workshop helped my professional development.
Q15	Since the Toolbox workshop, my own research worldview has shifted.
Q16	Since the Toolbox workshop, I have done additional reading on topics raised during the workshop.
Q17	Since the Toolbox workshop, I have discussed topics raised in the workshop with other group members present.
Q18	Since the Toolbox workshop, I have discussed topics raised in the workshop with others not in the workshop.

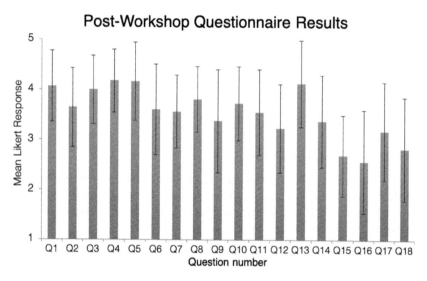

FIGURE 9.7 Summary of mean responses to the Toolbox post-workshop questionnaire by question.

While most of these results are encouraging, the responses to four questions stand out as worthy of comment. First, Toolbox workshops would not work without an open exchange between participants, so it is important that participants experience the dialogues as spaces in which thoughts move freely. It is therefore reassuring to see that participants report that they believed the dialogue was an open exchange of thoughts and ideas (Question 4, mean = 4.16). The second and third responses worth noting address the issue of group metacognition. Participants report that they have a better understanding of other group member's research because of the Toolbox dialogue (Question 9: mean = 3.38) and that the Toolbox workshop helps their groups identify some of their differing research worldviews (Question 10: mean = 3.73). One of the central goals of any Toolbox workshop is to help participants better understand their teammates' worldviews. The fact that participants themselves report enhanced mutual understanding further bolsters the results reported above. Finally, it is gratifying to see that many report that participating in a Toolbox workshop has led them to think that mutual understanding of differing research worldviews is important for productive collaboration in cross-disciplinary teams (Question 13: mean = 4.12). We want our workshops to raise awareness of what goes into successful collaboration, and this result suggests that they make progress toward this end.

9.6 CONCLUSION

Toolbox workshops are designed to foster integration. As noted in Chapter 5, integration within a collaborative project doesn't require team members to change their philosophical worldviews. It *does* require mutual understanding among team members of each other's philosophical worldviews. The results reported in this chapter demonstrate that Toolbox workshops are an effective means of fostering this mutual understanding. First, they strongly suggest that the dialogue during the workshop leads to the formation of beliefs about team members' worldviews, where before there was ignorance. Second, the results indicate that these beliefs about team members' worldviews increase in accuracy after the dialogue. These findings are supported by the testimony of participants themselves.

9.7 ACKNOWLEDGMENTS

We would like to thank Troy E. Hall for a helpful comment that was instrumental in the analysis of this data. An earlier version of this chapter was presented at the Public Philosophy Network 2018 Conference in Boulder, Colorado. We would like to thank the attendees for their useful feedback. Last, we thank the editors.

9.8 LITERATURE CITED

Bammer, G. 2013: Disciplining Interdisciplinarity: Integration and Implementation Sciences for Researching Complex Real-World Problems. Australian National University Press, Canberra.

Biasutti, M., and S. Frate. 2018: Group metacognition in online collaborative learning: Validity and reliability of the group metacognition scale (GMS). Educational Technology Research and Development 66:1321–1338.

Boulware-Gooden, R., S. Carreker, A. Thornhill, and R. M. Joshi. 2007: Instruction of meta-cognitive strategies enhances reading comprehension and vocabulary achievement of third-grade students. The Reading Teacher 61:70–77.

Brigandt, I. 2010: Beyond reduction and pluralism: Toward an epistemology of explanatory integration in biology. Erkenntnis 73:294–311.

———. 2013: Integration in biology: Philosophical perspectives on the dynamics of interdisciplinarity. Studies in History and Philosophy of Biological and Biomedical Sciences 44:461–465.

Chalmers, C. 2009: Group metacognition during mathematical problem solving. Pp. 1–8 *in* R. Hunter, T. Burgess, and B. Bicknell, eds. Crossing Divides: Proceedings of the 32nd Annual Conference of the Mathematics Education Research Group of Australasia. Mathematics Education Research Group of Australasia, Adelaide.

Cohen, J. 1988: Statistical Power Analysis for the Behavioral Sciences. 2nd edition. Erlbaum, Hillsdale, NJ.

Eigenbrode, S. D., M. O'Rourke, J. D. Wulfhorst, D. M. Althoff, C. S. Goldberg, K. Merrill, W. Morse, M. Nielsen-Pincus, J. Stephens, L. Winowiecki, and N. A. Bosque-Pérez. 2007: Employing philosophical dialogue in collaborative science. BioScience 57:55–64.

Flavell, J. H. 1979: Metacognition and cognitive monitoring: A new area of cognitive–developmental inquiry. American Psychologist 34:906–911.

Gerson, E. M. 2013: Integration of specialties: An institutional and organizational view. Studies in History and Philosophy of Science Part C: Studies in History and Philosophy of Biological and Biomedical Sciences 44:515–524.

Holm, S. 1979: A simple sequentially rejective multiple test procedure. Scandinavian Journal of Statistics 6:65–70.

Klein, J. T. 1990: Interdisciplinarity: History, Theory, and Practice. Wayne State University Press, Detroit.

———. 2012: Research integration: A comparative knowledge base. Pp. 283–298 *in* A. F. Repko, W. H. Newell, and R. Szostak, eds. Interdisciplinary Research: Case Studies of Integrative Understandings of Complex Problems. SAGE, Thousand Oaks, CA.

Looney, C., S. Donovan, M. O'Rourke, S. J. Crowley, S. D. Eigenbrode, L. Rotschy, N. A. Bosque-Pérez, and J. D. Wulfhorst. 2014: Seeing through the eyes of collaborators: Using Toolbox workshops to enhance cross-disciplinary communication. Pp. 220–243 *in* M. O'Rourke, S. J. Crowley, S. D. Eigenbrode, and J. D. Wulfhorst, eds. Enhancing Communication and Collaboration in Interdisciplinary Research. SAGE, Thousand Oaks, CA.

Metcalfe, J., and A. P. Shimamura. 1994: Metacognition: Knowing About Knowing. MIT Press, Cambridge, MA.

O'Rourke, M., and S. J. Crowley. 2013: Philosophical intervention and cross-disciplinary science: The story of the Toolbox Project. Synthese 190:1937–1954.

O'Rourke, M., S. J. Crowley, and C. Gonnerman. 2016: On the nature of cross-disciplinary integration: A philosophical framework. Studies in History and Philosophy of Science Part C: Studies in History and Philosophy of Biological and Biomedical Sciences 56:62–70.

Repko, A. F. 2007: Integrating interdisciplinarity: How the theories of common ground and cognitive interdisciplinarity are informing the debate on interdisciplinary integration. Issues in Integrative Studies 25:1–31.

Russell, B. 1997: The Problems of Philosophy. Oxford University Press, Oxford.

Siegel, M. A. 2012: Filling in the distance between us: Group metacognition during problem solving in a secondary education course. Journal of Science Education and Technology 21:325–341.

Teng, F. 2020: Tertiary-level students' English writing performance and metacognitive awareness: A group metacognitive support perspective. Scandinavian Journal of Educational Research 64: 551–568.

Teng, F., and B. L. Reynolds. 2019: Effects of individual and group metacognitive prompts on EFL reading comprehension and incidental vocabulary learning. PLOS ONE 14:e0215902.

Winne, P. H. 2017: Cognition and metacognition within self-regulated learning. Pp. 52–64 *in* D. H. Schunk and J. A. Greene, eds. Handbook of Self-Regulation of Learning and Performance. Routledge, New York.

10 Qualitative Analyses of the Effectiveness of Toolbox Dialogues

Marisa A. Rinkus
Michael O'Rourke

10.1 INTRODUCTION

The Toolbox Dialogue Initiative (TDI) has collected a wealth of data during more than a decade of studying and facilitating cross-disciplinary communication, including a large number of audio files and transcripts from the dialogue portion of Toolbox workshops. This book would not be complete without a discussion of these qualitative data. Building on the quantitative evidence of effectiveness proffered in Chapter 9, this chapter presents excerpts of Toolbox dialogue that illustrate how dialogue can enable teams to build mutual understanding, which can then lead to integration.

The centerpiece of TDI's work is dialogue-based workshops. Toolbox workshops aim to enhance mutual understanding and project integration by encouraging group reflexivity and perspective taking. In a Toolbox workshop, communication takes the form of "co-construction of meaning in pursuit of a goal" (Hall and O'Rourke 2014, p. 120; see also Chapter 5). This goal-directed co-construction often comes about through *dialogue*, a type of joint activity grounded in turn-taking that supports collective goal pursuit (see Chapter 7). This chapter focuses on the qualitative data collected during the dialogue portions of Toolbox workshops to illustrate how the presence of group reflexivity and perspective taking in dialogues can facilitate mutual understanding. Our analysis identifies emergent patterns in conversations that exhibit reflexivity and perspective taking, tracing how those contribute to the creation of mutual understanding and articulating the conditions under which mutual understanding supports integration within the dialogue.

We begin this chapter by presenting a conceptual framework for a process that relates the constructs *group reflexivity*, *perspective taking*, *mutual understanding*, and *project integration*, followed by introductions to these constructs. This chapter also contains excerpts of workshop dialogues that illustrate how dialogue can enable teams to build the sort of mutual understanding that can lead to integration. The bulk of the chapter is dedicated to analysis of this qualitative data set and a discussion of our findings.

10.2 CONCEPTUALIZING MUTUAL UNDERSTANDING CONVERSATIONS

We start by introducing the concept of "mutual understanding conversations," which are portions of the dialogue in which participants exhibit the sort of meaning making central to a successful Toolbox dialogue. Such conversations are anticipated in Chapter 7, where we follow Tsoukas (2009, p. 943) in taking dialogue to be a "joint activity between at least two speech partners, in which a turn-taking sequence of verbal messages is exchanged between them, aiming to fulfill a collective goal." Dialogue among partners who don't understand one another well enough to achieve their collective goal will need to build that understanding over the course of the exchange, and mutual understanding conversations are a critical part of such co-construction.

We view the process of building mutual understanding in Toolbox dialogue as driven by group reflexivity and perspective taking, which work in conjunction with each other to create the potential for mutual understanding conversations. Group reflexivity and perspective taking mutually reinforce each other. First, perspective taking across a group enhances the group's ability to be reflexive by situating each group member's points of view in the context provided by alternative perspectives. This process reveals and renders salient the distinguishing features of the various points of view, making each one more easily available for consideration by the group as a whole. Second, reflection on the different points of view in a group prepares members to take alternative perspectives by revealing common features among the perspectives, as well as differences that will require imagination to appreciate. The development of reflexivity and perspective taking among group members during a dialogue lays the groundwork for enhanced mutual understanding, thereby enabling collaborators to coordinate what they know and to better anticipate each other's project-relevant actions. Of course, laying the groundwork for mutual understanding does not guarantee that it will occur. It is possible to engage in group reflexivity and perspective taking in ways that are biased or in ways that exclude key aspects of a collaborator's views (cf. Settles et al. 2019). In these cases, efforts at mutual understanding will reinforce misunderstanding, driving into the background differences that must be acknowledged before real understanding can be achieved.

Integration of different perspectives, insights, and data into a coherent output is a primary goal for many groups with which TDI works (O'Rourke et al. 2016). Mutual understanding is an important integrated output in the context of cross-disciplinary collaboration. In a mutual understanding conversation, two collaborators transition from at least partial ignorance about one another's perspectives on a common problem to knowledge about those perspectives, facilitated often by reflection on their own contributions and heightened appreciation for how the contribution of the other could be beneficial. This epistemic transition is a form of integration because each party has expanded what they know about the perspectives held by members of their project. In doing so, they have related a previously unknown (or underappreciated) perspective to their own. More specifically, each party now has an understanding of the team's perspective, which combines the views of its members. In the best case, each party understands how to coordinate the perspectives of different collaborators. This coordination can take on different forms, such as *complementarization* if they cover different ground and *subordination* if one fits as a special case of the other. Once mutual understanding is achieved, it can support further integration of explanations, methods, or data (Brigandt 2013), which can also take place in the context of conversation. These types of integration are part of the common ground of the project, feeding back into mutual understanding in a way that promotes further integration.

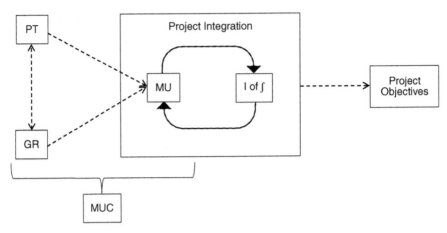

FIGURE 10.1 A conceptual framework for mutual understanding conversations (MUC). Here, PT = perspective taking, GR = group reflexivity, MU = mutual understanding, and I of ʃ = instances of integration. The dashed, double-headed arrow between group reflexivity and perspective taking represents these constructs as mutually reinforcing. Mutual understanding and instances of integration also reinforce one another, with the achievement of mutual understanding constituting and causing episodes of integrative activity in the project that can then create new opportunities for enhanced mutual understanding. We also theorize overall project integration as emerging from more local integrative episodes.

Over time, integrative moments in a project, including moments of enhanced mutual understanding, both cause and constitute whatever broader project integration is achieved, which can lead to research results and recommendations for addressing and solving problems.

In summary, mutual understanding conversations are conversations that support the integrative work of complex projects by way of the enhancement of mutual understanding. We believe that both group reflexivity and perspective taking can contribute to mutual understanding, and that mutual understanding supports and is supported by integration. Of course, perspective taking does not automatically produce reflexivity, and mutual understanding does not always lead to additional instances of integration. Furthermore, local instances of integration can lead to global project integration or feed back into mutual understanding conversations, creating an iterative process. Figure 10.1 presents our conceptual framework of the relationship and function of reflexivity, perspective taking, and mutual understanding in dialogue. We now consider each of these in turn.

10.3 REFLEXIVITY

Reflexivity is the process of introspective examination of one's own perspective. For an individual, reflexivity involves examining the assumptions, values, beliefs, social positions, and biases that influence one's judgments, decisions, and behaviors (Hesse-Biber 2014). Here we utilize the definition of "reflexivity" commonly operationalized as feminist research praxis in order to address and account for the influence of the researcher on the research itself and the subsequent data that are produced. At the level of the group or team, reflexivity is marked by the ability of group members to identify and examine the joint commitments that are key to the group's identity. At this level, reflexivity can be an important iterative process of discovery, adaptation, and action that facilitates group interaction and cohesiveness.

So understood, reflexivity is a key tool for working with others to complete a task or participate in an activity. This is emphasized by Salazar et al. (2019, p. 318), who define team reflexivity as "the explicit and purposeful reflection on the team's knowledge, strategy, and progress toward goals." Widmer et al. (2009, p. 2) also underscore the relationship between team reflexivity and action, noting that it corresponds to "the extent to which teams reflect upon and modify their functioning." Reflexivity can assist members of a team in learning who knows what (Lewis et al. 2007) and in knowledge coordination and sharing (West 1996), both of which enhance a team's effectiveness. The positive association between team reflexivity and team effectiveness indicates that reflexivity promotes team advancement (Widmer et al. 2009)

However, team reflexivity doesn't happen spontaneously. Certain conditions are conducive for it, such as collective communication competence, including the ability to listen to one another and manage disagreement without engendering ill will (cf. Thompson 2009; see Chapter 5). Collective communication competence can be cultivated by encouraging dialogue marked by deep listening and a commitment to the co-construction of meaning (Choi and Richards 2017). The Toolbox approach provides individuals and groups with a dialogical exercise in philosophical reflexivity by facilitating "awareness of the epistemic, metaphysical, and moral assumptions that shape one's research judgments, decisions, and behaviors" (Gonnerman et al. 2015, p. 674). This type of awareness can enable teams to enjoy the fruits of dialogue, which include critical self-reflection, enhanced awareness and appreciation of shared knowledge and of different assumptions, and the capacity to leverage epistemic diversity (see also Chapter 7).

10.4 PERSPECTIVE TAKING

The "act of taking another's perspective" involves an observer who "tries to understand, in a non-judgmental way, the thoughts, motives, and/or feelings of a target, as well as why they think and/or feel the way they do" (Parker et al. 2008, p. 151). This can be described as seeing things "through someone else's eyes" (cf. Looney et al. 2014). It is an intentional activity that can be thought of as using your imagination to simulate the perspective of another person (Harris 1996; Farrant et al. 2006). Although it may be possible to imagine the perspective of one you know well without too much effort, the active acquisition of information is typically required when the target of one's perspective taking is less well-known. Characterized in this way, perspective taking is closely related to perspective *seeking*, which involves "actively seeking out the viewpoints" of others (Salazar et al. 2019, p. 319). Perspective taking will often be achieved as a consequence of perspective seeking, which is itself an intentional and goal-directed activity (Parker et al. 2008).

Perspective taking by team members can be viewed as a mechanism that enables creativity (Hoever et al. 2012), effective communication (Gockel and Brauner 2013), and integrative thinking (Grant and Berry 2011). Hoever et al. argue (p. 982) that perspective taking facilitates information elaboration, requiring team members to "invest cognitive energy in understanding" diverse perspectives and approaches through constructive discussion and integration of the viewpoints of other team members. Seeking the perspective of others through structured group dialogue has been found to aid in knowledge integration by shifting the focus from self to others and by stimulating interaction (Okhuysen and Eisenhardt 2002). Perspective taking can be employed as a strategy for obtaining "more accurate knowledge about others' knowledge" that can be used in multiple situations (Gockel and Brauner 2013, p. 229).

It often takes work to understand how another person thinks and feels about a topic (cf. Ruby and Decety 2001). One must try to understand their actions and reactions and also be sensitive to feedback that indicates the success of one's effort at perspective taking. It is important to guard against the adoption of stereotypes that can result in less accurate understanding of how another thinks and feels (Galinsky et al. 2008). One way to work toward informed, critical perspective taking is to engage in dialogue and articulate the core beliefs and values that frame one's perspective. Dialogue of the sort found in Toolbox workshops can help teammates come to understand the beliefs and values that their collaborators bring to their common research project (O'Rourke and Crowley 2013; Looney et al. 2014). For instance, dialogue provides a mechanism for evaluating the degree to which collaborators understand one another's viewpoints. Toolbox workshops offer a context in which collaborators can articulate their perspectives and thereby supply their teammates with information they can then use for informed perspective taking.

10.5 MUTUAL UNDERSTANDING AND CREATING COMMON GROUND

We view mutual understanding as an achievement gained from team reflection and perspective taking. Mutual understanding is exhibited by a group in which members have systematic knowledge of one another. Several observations are important here. First, mutual understanding is more robust if it goes beyond simple knowledge of one another. In most circumstances, group members have a minimal degree of mutual understanding that involves facts about each other (e.g., that one team member is a biologist, another team member is an engineer, and still another went to Michigan State), but more robust mutual understanding involves some degree of empathetic appreciation for how each other *finds* the world, i.e., what a team member cares about and knows. Second, more robust mutual understanding is grounded in *common* or *mutual* knowledge (Lewis 1969; Schiffer 1972). Mutual knowledge exists if, for example, the biologist knows something about the engineer, the engineer knows that the biologist knows this, the biologist knows that the engineer knows this about the biologist, and so on. Common knowledge of this sort knits a group together into a collective, ensuring that people share knowledge about one another openly and can coordinate their actions on that basis. Third, mutual understanding entails *self-*understanding. A team with high mutual understanding includes teammates who are self-aware and capable of representing their perspectives in discourse (Gonnerman et al. 2015; cf. Morin 2011). In groups with members who are new to one another, conversation is an important mechanism for generating mutual understanding (Ta et al. 2017), in part because each individual communicates their own perspective and thereby helps others understand them (Berger and Calabrese 1974).

In the literature on communication and interdisciplinarity, mutual understanding is closely associated with *common ground*. Stalnaker (1978. p. 321) defines "common ground" to be "the common knowledge or mutual knowledge" of speakers in a conversation. In defining it this way, he draws on Lewis (1969) and Schiffer (1972) and also Grice (1989), whose notion of "common ground status" is the original source of this framework. (Stalnaker 2002 expands on this attribution to Grice.) Clark (1996, p. 93) adds joint beliefs and suppositions to common ground, along with knowledge. Bromme (2000, p. 119) summarizes Clark's conception of common ground as "a common cognitive frame of reference between the partners" of a communicative interaction, and that "all contributions to the process of mutual understanding serve to establish or ascertain and continually maintain this common ground." Thus, the dynamic common ground of a conversation can be regarded as the output of the process of mutual understanding. For a stable group that participates in much conversation, common ground can grow and become a shared basis that supports the anticipation, coordination, and facilitation that mark high-functioning teams (Quinlan and Robertson 2010; cf. Clark 1996).

As described in previous chapters, communication in cross-disciplinary contexts is fraught with difficulties for a variety of reasons. One key problem can be the lack of a robust common ground among collaborators. Representatives of different disciplines are trained in different ways, learn different methods and techniques, and become fluent in different technical languages. While all team members may be experts, this doesn't ensure that they have the ability to communicate their perspective to those who look at things very differently than they do. Building common ground that can support communication and coordination among cross-disciplinary collaborators is key to developing integrated understanding of their common project. One form of mutual understanding that matters significantly in collaborative cross-disciplinary settings is knowledge of what facts each collaborator knows, what research problems animate them, what methods they apply, and so forth. In a cross-disciplinary research environment, Repko (2012) argues that this sort of common ground is created by sharing, negotiating, and modifying concepts, theories, or issues in order to co-create knowledge. Opportunities to engage in collective reflection and perspective taking in conversation about research styles, such as Toolbox workshops, are especially potent contexts for the development of this sort of common ground. They build the mutual understanding necessary for successful pursuit of integrative research objectives.

10.6 TOOLBOX QUALITATIVE DATA

Most of the qualitative data that TDI collects derive from the dialogue portions of Toolbox workshops. There are over 200 hours of recorded dialogue from over 150 Toolbox workshops. Approximately 50% of the recorded dialogues have been transcribed and have had their statements linked to individual speakers based on speaking turn notes taken by an observer. Here, we present our findings from six workshops that used the Scientific Research Toolbox instrument, which focuses on foundational aspects of scientific research such as the nature of scientific knowledge and the nature of the investigated world (O'Rourke and Crowley 2013; Looney et al. 2014; see Appendix A). These workshops involved a heterogeneous set of teams that varied in terms of context (administration, ad hoc, research), composition (team, non-team), size (small to large), career stage (graduate student to faculty member), and discipline (see Table 10.1). Five of the teams were assembled for the purpose of research and one was an ad hoc group assembled in advance of a conference

TABLE 10.1
Characteristics of the Dialogue Sessions

Dialogue Session	Group Description	Number of Participants	Relationship	Dialogue Duration
DS1	The governing body for an academic center focused on sustainability; its purpose was to make leadership decisions for the center and determine allocation of research support resources.	6 (Faculty)	Team	72 min
DS2	Members of an Integrative Graduate Education and Research Traineeship (IGERT) project created to work on biofuels, plant-made products, and environmental sustainability.	4 (Faculty and students)	Team	70 min
DS3	A university group interested in submitting a proposal to the National Science Foundation (NSF).	12 (Faculty)	Team	104 min
DS4	A team of graduate students who participated in an IGERT project; the workshop was part of the IGERT launch meeting, and was the first meeting for this team.	3 (Students)	Team	97 min
DS5	A team of graduate students who participated in an IGERT project; the workshop was part of the launch meeting, and was the first meeting for this team.	4 (Students)	Team	89 min
DS6	Attendees at a workshop aimed at developing and sustaining interdisciplinary graduate programs.	10 (Faculty)	Non-team (ad hoc group)	70 min

that focused on graduate education. Of the five teams, three were small research teams, a team planning an NSF grant proposal, and a team of administrators that ran a research center. The team/non-team distinction is an important difference among the dialogue groups because team members generally know one another better than non-teammates and because teammates with collective goals have additional motivation to communicate openly and constructively.

Group size for the dialogue sessions ranged from 3 to 12 participants (average of approximately 6 participants). Three of the workshops were conducted with faculty only, two were conducted with graduate students, and one included a graduate student, a post-doctoral researcher, and two faculty. Participants represented various disciplines including biology, ecology, economics, engineering, entomology, hydrology physics, sociology, and soil science. The length of the dialogue session in the workshops ranged from 70 to 104 minutes.

10.7 ANALYSIS

The transcripts from the six workshops presented here were analyzed using a coding structure developed for a larger study; only the aspects that relate to the goals of this chapter are discussed in this section. The dialogues were transcribed verbatim and the function and impact of each speaking turn in the dialogue was coded. As we will see, a speaking turn could have multiple functions or impacts on the dialogue, and therefore multiple codes. An example of a function code is "query," which denotes a participant asking a question of an individual or the group in order to seek their perspective or for clarification of a perspective. The function codes we focus on as representative of perspective taking include "elaborate," "example," "query," and "challenge."

Impact codes denote a particular bearing on the dialogue or participant(s). For example, the impact code "self-aware" represents reflexivity and is assigned to a speaking turn that manifests a speaker's developing self-awareness of the issues under discussion. The code "team aware" is assigned when there is evidence of collective awareness of differences or similarities within the team concerning the issues under discussion. Two other impact codes of interest to us here are "co-create: construct" and "co-create: negotiate." These indicate mutual understanding and common ground. Mutual understanding conversations (MUCs) are sections of the dialogue (i.e., multiple speaking turns) in which participants exhibit the meaning making that is the goal of the Toolbox dialogue. We consider the exchanges that are coded as "co-create: construct" and "co-create: negotiate" to be paradigmatic MUCs. Table 10.2 illustrates the relationship between key outcomes of the dialogue and the function and impact codes.

10.8 MUTUAL UNDERSTANDING, REFLEXIVITY, AND PERSPECTIVE TAKING IN TOOLBOX DIALOGUE

In this section we present excerpts from six Toolbox workshop dialogues that provide evidence of reflexivity, perspective taking, and mutual understanding. These excerpts highlight important moments in dialogue. Each excerpt starts with

TABLE 10.2
Key Characteristics and Corresponding Codes with Descriptions

Key Outcomes	Function and Impact Codes
Reflexivity—individual and group	Self Aware—developed self-awareness of issues Self Shift—shift in commitments concerning the issues Team Aware—airing of awareness of differences or similarities regarding issues discussed by the team
Perspective taking	Info—introduces new information about issues Query—asks a question of an individual or the group Example—provides a concrete example Elaborate—elaborates on the content of a statement made Challenge—speaker states a position that is different than one previously described
Mutual understanding	Co-create: Construct—co-creation of shared meaning from elaboration/questioning that leads to new points or differentiation of prior points Co-create: Negotiate—co-creation of meaning from disagreement accompanied by discussion about disagreement

a quotation that exemplifies some aspect of the key outcomes of dialogue and is followed by a narrative that includes our analysis and interpretation. To ensure confidentiality, all participants are referred to by participant number (e.g., P1, P2). The excerpts are annotated with our coding scheme in order to highlight the characteristics of the dialogue that exhibit aspects of reflexivity, perspective taking, and mutual understanding.

10.8.1   Excerpt 1: "Yeah, It's Completely Relative to What We're Used to Doing. That Is Interesting ...."

This excerpt illustrates how perspective taking and reflexivity work together to support mutual understanding through queries and examples that enhance self and team awareness of basic and applied research (See Figure 10.2). The discussion is driven by the prompt "*The importance of our project stems from its applied aspects*" from the Motivation module of the Scientific Research Toolbox instrument. Three of the four participants from this team have an engaged dialogue about whether their team project is applied or basic (ST 37–52) and assert that their project is more applied than their disciplinary research (ST 38) and more basic than industry research (ST 45). (ST denotes speaking turn.)

The introduction of examples and queries, such as ST 45 in which P1 introduces the idea of *application* from industry, leads to additional queries, examples, and validation that result in mutual understanding within the group. Although the motivation for the team's research is a "product" (i.e., an application), there are fundamental questions along the way (e.g., P3's comments in ST 40, ST 42, ST 44). There is

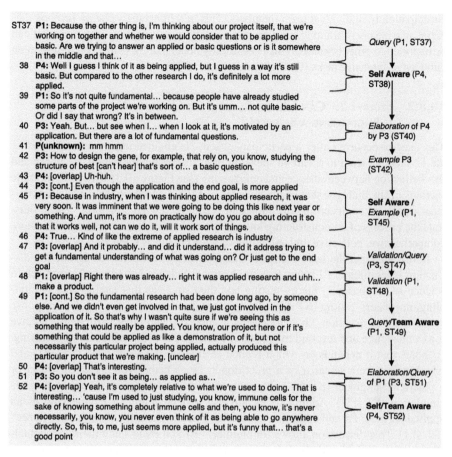

FIGURE 10.2 Excerpt from dialogue session DS2 focusing on the prompt, "The importance of our project stems from its applied aspects." We interpret "fundamental" and "basic" as synonymous for the purposes of this exchange. (Function codes are *italicized*; impact codes are in **boldface**.)

evidence of self- and team-awareness in ST 52 in which P4 acknowledges that understanding basic vs. applied research is relative to what one is used to doing. In affirming this point, P4 endorses this as an insight into their team's work. This excerpt represents an important discussion for the team as they situate their project within their own disciplinary research and establish common ground relative to applied and basic research.

In this example, we see perspective taking as an "emergent team process" (Hoever et al. 2012) that stimulates thinking about individual viewpoints on basic and applied research and leads the group to discuss their own project. The discussion reveals multiple perspectives that could contribute to the conceptualization of the project and also contribute to the team's understanding of the different motivations for participation by individual team members. This highlights the value of Toolbox dialogue as a means of training team members in perspective-taking strategies that can be exported to other contexts. This is especially beneficial for teams collaborating on knowledge-intensive tasks (Gockel and Brauner 2013).

10.8.2   EXCERPT 2: "CAN I ASK A REALLY SILLY QUESTION? IS ENGINEERING A SCIENCE?"

This excerpt occurred after a participant asked, "Is engineering a science?" leading the 12-member team to spend 46 speaking turns discussing the nature of science and the distinction between science and engineering (see Figure 10.3). There are multiple sections of the dialogue that show evidence of MUCs, in which the participants discuss aspects of engineering education and academic reward structures. These illustrate the process of co-constructing knowledge as they work their way toward a more comprehensive, collective understanding of what counts as science.

In this passage, the team members discuss whether the science/engineering distinction is legitimate. P1 references the National Science Foundation's approach to science as emphasizing theory testing and then suggests that engineering is modeling, not theory testing (ST 448). P1 initiates development of a negative response to the opening question, "Is engineering a science?" and other participants receive it as a challenge and respond by presenting alternative perspectives on engineering. P7 notes that it depends on which branch of engineering is involved (ST 451) and whether you are a scientist or a practitioner, i.e., whether you are a researcher or a practicing engineer (ST 454). P9 notes that modeling permits theory testing via simulation (ST 457). Adding to the diverse understanding of engineering as an applied science, P5 adds that the "core of engineering is a lot of physical science" (ST 461). This discussion allows for engineers P7 and P9 to share their conceptualization of their field with the larger team and negotiate what that means in the team context. In seeking the perspective of the engineers on the team, P1 and other non-engineer team members increase their understanding and awareness of engineering as a field and, more importantly, how the engineers on the team position themselves as researchers. Although there isn't a straightforward answer to the question of whether engineering is a science, the team has been primed to think about it as important.

This excerpt illustrates a reflexive attempt by the team to learn more about how to think of engineering. It began with perspective taking, which is significant because the research team was an interdisciplinary group that included engineers. The dialogue here was enabled by the Toolbox process, which allowed this question to emerge organically in the course of dialogue and be asked openly and considered in a safe context. Since questions of identity and worldview are main emphases of the dialogue, working out similarities and differences at the level of fundamental research orientation is a way of assessing your collaborators and determining how to relate to each other. This is a key element of perspective taking, which often requires gathering information that can be used to understand your project as your collaborators do (Parker et al. 2008; Hoever et al. 2012).

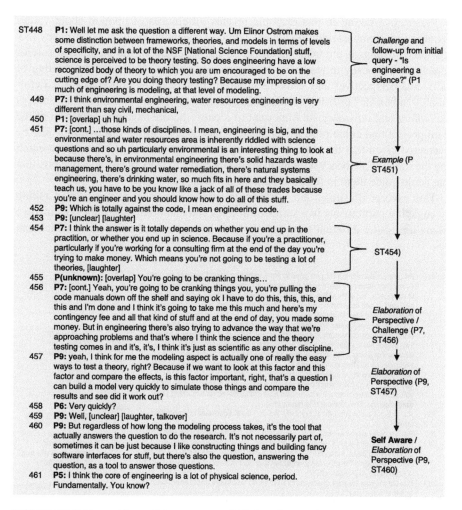

ST448 P1: Well let me ask the question a different way. Um Elinor Ostrom makes some distinction between frameworks, theories, and models in terms of levels of specificity, and in a lot of the NSF [National Science Foundation] stuff, science is perceived to be theory testing. So does engineering have a low recognized body of theory to which you are um encouraged to be on the cutting edge of? Are you doing theory testing? Because my impression of so much of engineering is modeling, at that level of modeling.

Challenge and follow-up from initial query - "Is engineering a science?" (P1

449 P7: I think environmental engineering, water resources engineering is very different than say civil, mechanical,

450 P1: [overlap] uh huh

451 P7: [cont.] ...those kinds of disciplines. I mean, engineering is big, and the environmental and water resources area is inherently riddled with science questions and so uh particularly environmental is an interesting thing to look at because there's, in environmental engineering there's solid hazards waste management, there's ground water remediation, there's natural systems engineering, there's drinking water, so much fits in here and they basically teach us, you have to be you know like a jack of all of these trades because you're an engineer and you should know how to do all of this stuff.

Example (P ST451)

452 P9: Which is totally against the code, I mean engineering code.

453 P9: [unclear] [laughter]

454 P7: I think the answer is it totally depends on whether you end up in the practition, or whether you end up in science. Because if you're a practitioner, particularly if you're working for a consulting firm at the end of the day you're trying to make money. Which means you're not going to be testing a lot of theories, [laughter]

ST454)

455 P(unknown): [overlap] You're going to be cranking things...

456 P7: [cont.] Yeah, you're going to be cranking things you, you're pulling the code manuals down off the shelf and saying ok I have to do this, this, this, and this and I'm done and I think it's going to take me this much and here's my contingency fee and all that kind of stuff and at the end of day, you made some money. But in engineering there's also trying to advance the way that we're approaching problems and that's where I think the science and the theory testing comes in and it's, it's, I think it's just as scientific as any other discipline.

Elaboration of Perspective / Challenge (P7, ST456)

457 P9: yeah, I think for me the modeling aspect is actually one of really the easy ways to test a theory, right? Because if we want to look at this factor and this factor and compare the effects, is this factor important, right, that's a question I can build a model very quickly to simulate those things and compare the results and see did it work out?

Elaboration of Perspective (P9, ST457)

458 P6: Very quickly?

459 P9: Well, [unclear] [laughter, talkover]

460 P9: But regardless of how long the modeling process takes, it's the tool that actually answers the question to do the research. It's not necessarily part of, sometimes it can be just because I like constructing things and building fancy software interfaces for stuff, but there's also the question, answering the question, as a tool to answer those questions.

Self Aware / *Elaboration* of Perspective (P9, ST460)

461 P5: I think the core of engineering is a lot of physical science, period. Fundamentally. You know?

FIGURE 10.3 Excerpt from dialogue session DS3 focusing on the question, "Is engineering a science?" raised in the dialogue by one of the participants. (Function codes are *italicized*; impact codes are in **boldface**.)

10.8.3 Excerpt 3: "I've Got a Question for Everybody. Do You Think Advocacy Is a Negative Term?"

This excerpt occurred when a team of four graduate students discussed three prompts relating to values from the Toolbox instrument. The portion presented here is their discussion of advocacy and scientific research related to the prompt, *"Allowing values to influence scientific research is advocacy"* (see Figure 10.4). The discussion illustrates how a workshop can support MUCs through dialogue that enables perspective taking by disclosing differences of opinion about values in the team, and fosters reflexivity by generating self and team awareness.

Perspective taking in this example reveals differences of opinion about values among team members and initiates an explicit conceptual analysis of advocacy (ST 359–363). Although the team members do not agree about the nature and role of values and advocacy in research, the discussion produces deeper understanding of individual perspectives and leads the group to appreciate the importance of a definition of "advocate" (ST 365). Differences in opinion and understanding of concepts like *values* and *advocacy* are important to identify before engaging an aspect of the team project that could raise the prospect of advocacy. Further discussion would be warranted given that the participants disagreed about the role of values in research and the status of advocacy without having the chance to work out a collective position that could guide them in future situations involving advocacy.

This excerpt illustrates that challenges can be important to perspective taking by engaging participants in a series of queries and elaborations. Challenges involve taking a position different from one that has been previously presented, which can

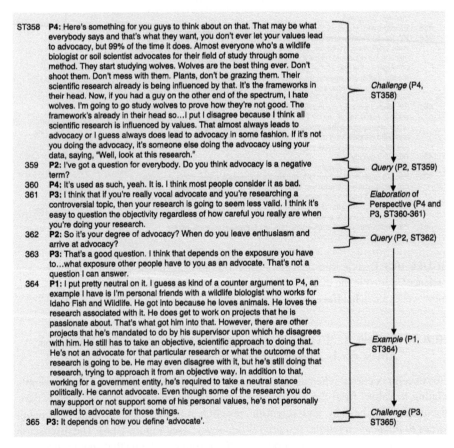

FIGURE 10.4 Excerpt from dialogue session DS5 focusing on the prompt, "Allowing values to influence scientific research is advocacy." (Function codes are *italicized*.)

produce uncertainty about the reception of the new perspective. This involves some risk on the part of the challenger. When done in a structured and well-facilitated environment that welcomes disagreement, challenges can ensure more probing consideration of the issue at hand. Exchanges that include challenges can develop mutual understanding through the co-creation and negotiation of knowledge. In the passage, the challenge occurs when P4 makes specific claims regarding values and advocacy. The workshop participants engage with the challenge by asking questions, providing their own perspective, and making counter arguments (ST 364) in a manner that can be considered as productive disagreement and negotiation in support of mutual understanding.

10.8.4   EXCERPT 4: "WELL, EXPLAIN MY POSITION. I'M NOT EVEN SURE WHAT IT IS."

The dominant theme of the first half of this dialogue session is what constitutes science. Discussion of this topic stems from the core questions and prompts in the instrument that concern confirmation (e.g., "*What types of evidentiary support are required for knowledge?*") (see Figure 10.5). However, much of the discussion of this topic diverges from the prompts. This commonly happens with teams that comprise individuals who have significant experience with the issues under discussion and thereby are comfortable going "off book." Prior to the excerpt, the team discussed whether or not something is considered science if you can't run an experiment, and also the role of replication in science. P2 asserts that non-experimental work is "*in the realm of scholarly activity*" but "*wouldn't call it science*" and further, that "*scholarly activity is not all research.*" This leads us to the excerpt presented here, in which participants employ elements of perspective taking in the form of challenges, elaborations, and queries as they work toward mutual understanding.

This dialogue contains negotiations about what counts as science. It begins when P1 expresses difficulty with P2's position that some research is "scholarly" and not "scientific," which prompts P2 to challenge P1 to explain their position (ST 263–264). This is an explicit instance of perspective *seeking* in that P2 seeks to understand P1's interpretation of their perspective. Understanding how others perceive what we say is an important aspect of individual reflexivity. It is also key to group reflexivity, because mutual understanding among team members will typically be necessary to support effective deliberation and decision making.

At the end of this excerpt, P2 connects the discussion of whether or not something is science to the work this team is doing in an academic center and what this means in the context of funding research efforts (ST 271). This conversation exposes deep views about what counts as science and how it should be fostered, which is an important goal of the center. It demonstrates how greater mutual understanding that arises from team reflexivity and perspective taking can inform how important project decisions are framed and engaged in by the group. Knowing the core beliefs and values of your collaborators allows you to avoid rhetoric that can obstruct collective consideration of important project issues.

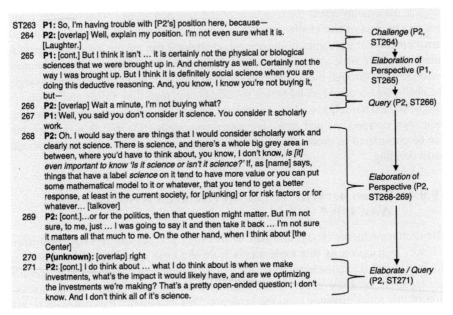

ST263 **P1:** So, I'm having trouble with [P2's] position here, because—
264 **P2:** [overlap] Well, explain my position. I'm not even sure what it is. [Laughter.]
265 **P1:** [cont.] But I think it isn't … it is certainly not the physical or biological sciences that we were brought up in. And chemistry as well. Certainly not the way I was brought up. But I think it is definitely social science when you are doing this deductive reasoning. And, you know, I know you're not buying it, but—
266 **P2:** [overlap] Wait a minute, I'm not buying what?
267 **P1:** Well, you said you don't consider it science. You consider it scholarly work.
268 **P2:** Oh. I would say there are things that I would consider scholarly work and clearly not science. There is science, and there's a whole big grey area in between, where you'd have to think about, you know, I don't know, *is [it] even important to know 'Is it science or isn't it science?'* If, as [name] says, things that have a label *science* on it tend to have more value or you can put some mathematical model to it or whatever, that you tend to get a better response, at least in the current society, for [plunking] or for risk factors or for whatever… [talkover]
269 **P2:** [cont.]…or for the politics, then that question might matter. But I'm not sure, to me, just … I was going to say it and then take it back … I'm not sure it matters all that much to me. On the other hand, when I think about [the Center]
270 **P(unknown):** [overlap] right
271 **P2:** [cont.] I do think about … what I do think about is when we make investments, what's the impact it would likely have, and are we optimizing the investments we're making? That's a pretty open-ended question; I don't know. And I don't think all of it's science.

Challenge (P2, ST264)

Elaboration of Perspective (P1, ST265)

Query (P2, ST266)

Elaboration of Perspective (P2, ST268-269)

Elaborate / Query (P2, ST271)

FIGURE 10.5 Excerpt from dialogue session DS1 focusing on two prompts: "Validation of evidence requires replication," and "Unreplicated results can be validated if confirmed by a combination of several different methods." (Function codes are *italicized*.)

10.8.5 Excerpt 5: "If You're Going to Work on a Team with People Who Don't Share Your Opinion, You at Least Have to Have the Conversation about How Everybody Is Constructing That."

This excerpt demonstrates the kind of probing and revealing dialogue a Toolbox workshop can create for groups that are not teams (see Figure 10.6). This dialogue occurred in an ad hoc group of 10 faculty members. The participants discussed how they interpreted the prompts differently, thereby allowing the group to air differences and consider multiple perspectives. Use of the numerical ratings in the Toolbox instrument to express individual-level agreement (ST 220–221) supports reflexivity by allowing the individual to reflect on their own perspective while also informing perspective taking by disclosing any potential disagreement (Rinkus et al. 2020). The excerpt is a MUC that occurred during a larger discussion of the prompt "*Scientific research aims to identify facts about a world independent of the investigators.*" It illustrates the mutually reinforcing nature of reflexivity and perspective taking.

Much of the discussion here focuses explicitly on terminology, including the meaning of terms such as "facts," "knowledge," "research," "process," and "mechanism." P2 distinguishes between research that is about "facts" and research that is about "meaning" or "experience" (ST 216). After a query by P1 about whether P2 would endorse the prompt if it used "knowledge" instead of "facts," P2 concludes that "different kinds of work from this spectrum can co-exist, hopefully peacefully." P1 takes the discussion a step further by introducing the topic of "prediction" as a

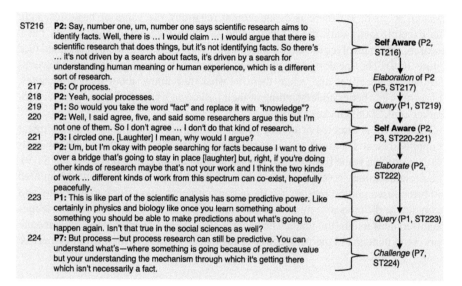

ST216 **P2:** Say, number one, um, number one says scientific research aims to identify facts. Well, there is ... I would claim ... I would argue that there is scientific research that does things, but it's not identifying facts. So there's ... it's not driven by a search about facts, it's driven by a search for understanding human meaning or human experience, which is a different sort of research. — **Self Aware** (P2, ST216)

Elaboration of P2 (P5, ST217)

217 **P5:** Or process.
218 **P2:** Yeah, social processes.
219 **P1:** So would you take the word "fact" and replace it with "knowledge"? — *Query* (P1, ST219)
220 **P2:** Well, I said agree, five, and said some researchers argue this but I'm not one of them. So I don't agree ... I don't do that kind of research. — **Self Aware** (P2, P3, ST220-221)
221 **P3:** I circled one. [Laughter] I mean, why would I argue?
222 **P2:** Um, but I'm okay with people searching for facts because I want to drive over a bridge that's going to stay in place [laughter] but, right, if you're doing other kinds of research maybe that's not your work and I think the two kinds of work ... different kinds of work from this spectrum can co-exist, hopefully peacefully. — *Elaborate* (P2, ST222)
223 **P1:** This is like part of the scientific analysis has some predictive power. Like certainly in physics and biology like once you learn something about something you should be able to make predictions about what's going to happen again. Isn't that true in the social sciences as well? — *Query* (P1, ST223)
224 **P7:** But process—but process research can still be predictive. You can understand what's—where something is going because of predictive value but your understanding the mechanism through which it's getting there which isn't necessarily a fact. — *Challenge* (P7, ST224)

FIGURE 10.6 Excerpt from dialogue session DS6 focusing on the prompt, "Scientific research aims to identify facts about a world independent of the investigators." (Function codes are *italicized*; impact codes are in **boldface**.)

hallmark of scientific analysis (ST 223), leading to an exchange between P1 and P7 that highlights P1's positivist approach to science in contrast with the approaches of P2, P3, P5, and P7. The final exchange between P1 and P7 results in a challenge by P7, who takes P1 to assume that facts underlie predictions and suggests that processes, which may not count as facts, could also ground predictions.

The dialogue from which this excerpt is taken contains further queries, examples, and challenges (see Appendix B). These moves in the dialogue help to reveal the underlying commitments that influence each individual's research worldview. This group exhibits some fundamental differences and, were it a team, a discussion of this nature could prove crucial to the creation of the mutual understanding necessary to continue their research project. The dialogue appears to create awareness among the participants that these types of discussions are necessary when working on a team, as exemplified by the introductory quote. Even outside of the team setting this excerpt is evidence of how Toolbox workshops encourage co-construction and negotiation of meaning through reflexivity and perspective taking.

10.8.6 EXCERPT 6: "SOMETHING THAT I HAVE PICKED UP ON ALREADY IS DEFINITELY YOUR BACKGROUND YOU'RE COMING FROM ... AND I JUST REALIZED IT WAS DIFFERENT BETWEEN US ...."

This team immediately addresses key issues related to their collective work together. The first four speaking turns in this dialogue (not shown) are the team members stating their initial reactions to the prompt, *"The principal value of research stems from the potential application of the knowledge gained."* All agree that applied research is good, but there is disagreement regarding the value of basic research. The excerpt opens with the second round of discussion about applied and basic research, which involves the elaboration of the perspectives previously presented (see Figure 10.7). The speaking turns here are longer as the participants work to articulate their perspectives and relate them to what they know about their colleagues. This results in a better understanding of their colleague's knowledge and perspectives (Parker et al. 2008; Gockel and Brauner 2013).

The excerpt is directed primarily by P2, who clarifies and affirms a commitment to applied research (wants to "*help the most people*," ST 5). P1 and P3, by contrast, accept that some basic research without application is acceptable (ST 6, ST 9). These contributions reveal self-awareness of the different commitments to applied and basic research, the reasons for these commitments, and efforts to negotiate and reconcile the differences among them. This is evident in ST 8 when P2 acknowledges the differences within the team. The excerpt also offers a clear example of someone working to understand the perspective of another in relation to their own perspective

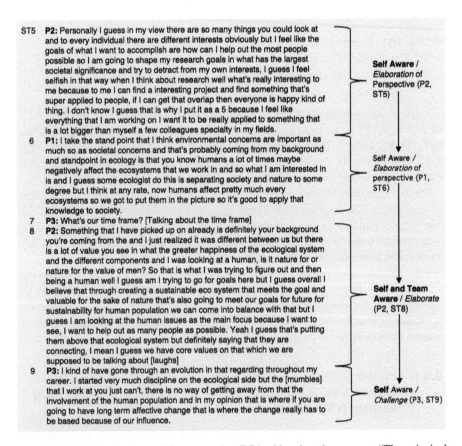

FIGURE 10.7 Excerpt from dialogue session DS4 addressing the prompt, "The principal value of research stems from the potential application of the knowledge gained." (Function codes are *italicized*; impact codes are in **boldface**.)

through the process of dialogue. P3 describes the evolution of their thinking and then challenges the team to consider the stance that humans must be considered in eco-logical research (ST 9). The reflexive comments here underscore the role of group reflexivity in building a foundation for mutual understanding.

10.9 CONCLUSION

We have used Toolbox transcripts to provide examples of group reflexivity, per-spective taking, and mutual understanding in Toolbox dialogues. The MUCs in the excerpts illustrate only a few of the many ways in which structured dialogue helps collaborators build mutual understanding through reflexivity and perspec-tive taking. Our analysis highlights the multiple aspects of group reflexivity, per-spective taking, and mutual understanding. These findings support our view that perspective taking and group reflexivity are mutually reinforcing in their facilitation of mutual understanding, and our view that MUCs are very beneficial for teams and ad hoc groups. Further research into the nature and roles of MUCs within a Toolbox dialogue in fostering integration is needed.

10.10 ACKNOWLEDGMENTS

We thank Troy E. Hall for leading the effort to produce a code book for Toolbox dia-logue and for contributing to the coding of the transcripts discussed in this chapter, and Chad Gonnerman for his work on MUCs. We also thank Graham Hubbs and Steven Orzack for their editing assistance. O'Rourke's work on this chapter was supported by the USDA National Institute of Food and Agriculture, Hatch Project 1016959.

10.11 LITERATURE CITED

Berger, C. R., and R. J. Calabrese. 1974: Some explorations in initial interaction and beyond: Toward a developmental theory of interpersonal communication. Human Communication Research 1:99–112.

Brigandt, I. 2013: Integration in biology: Philosophical perspectives on the dynamics of interdisciplinarity. Studies in History and Philosophy of Biological and Biomedical Sciences 44:461–465.

Bromme, R. 2000: Beyond one's own perspective: The psychology of cognitive interdisciplinarity. Pp. 115–133 *in* N. Stehr and P. Weingart, eds. Practicing Interdisciplinarity. University of Toronto Press, Toronto.

Choi, S., and K. Richards. 2017: Interdisciplinary Discourse. Palgrave Macmillan, London.

Clark, H. H. 1996: Using Language. Cambridge University Press, Cambridge.

Farrant, B. M., J. Fletcher, and M. T. Maybery. 2006: Specific language impairment, theory of mind, and visual perspective taking: Evidence for simulation theory and the develop-mental role of language. Child Development 77:1842–1853.

Galinsky, A. D., C. S. Wang, and G. Ku. 2008: Perspective-takers behave more stereotypically. Journal of Personality and Social Psychology 95:404–419.

Gockel, C., and E. Brauner. 2013: The benefits of stepping into others' shoes: Perspective taking strengthens transactive memory. Basic and Applied Social Psychology 35:222–230.

Gonnerman, C., M. O'Rourke, S. J. Crowley, and T. E. Hall. 2015: Discovering philosophical assumptions that guide action research: The reflexive toolbox approach. Pp. 673–680 *in* H. B. Huang and P. Reason, eds. The SAGE Handbook of Action Research, 3rd edition SAGE, Thousand Oaks, CA.

Grant, A. M., and J. W. Berry. 2011: The necessity of others is the mother of invention. Academy of Management Journal 54:73–96.

Grice, H. P. 1989: Studies in the Way of Words. Harvard University Press, Cambridge, MA.

Hall, T. E., and M. O'Rourke. 2014: Responding to communication challenges in transdisciplinary sustainability science. Pp. 119–139 *in* K. Huutoniemi and P. Tapio, eds. Heuristics for Transdisciplinary Sustainability Studies: Solution-Oriented Approaches to Complex Problems. Routledge, Oxford.

Harris, P. L. 1996: Desires, beliefs, and language. Pp. 200–220 *in* P. Carruthers and P. K. Smith, eds. Theories of Theories of Mind. Cambridge University Press, Cambridge.

Hesse-Biber, S. N. 2014: Feminist Research Practice: A Primer. SAGE, Thousand Oaks, CA.

Hoever, I. J., D. Van Knippenberg, W. P. Van Ginkel, and H. G. Barkema. 2012: Fostering team creativity: Perspective taking as key to unlocking diversity's potential. Journal of Applied Psychology 97:982–996.

Lewis, D. K. 1969: Convention. A Philosophical Study. Harvard University Press, Cambridge, MA.

Lewis, K., M. Belliveau, B. Herndon, and J. Keller. 2007: Group cognition, membership change, and performance: Investigating the benefits and detriments of collective knowledge. Organizational Behavior and Human Decision Processes 103:159–178.

Looney, C., S. Donovan, M. O'Rourke, S. J. Crowley, S. D. Eigenbrode, L. Rotschy, N. A. Bosque-Pérez, and J. D. Wulfhorst. 2014: Seeing through the eyes of collaborators: Using Toolbox workshops to enhance cross-disciplinary communication. Pp. 220–243 *in* M. O'Rourke, S. J. Crowley, S. D. Eigenbrode, and J. D. Wulfhorst, eds. Enhancing Communication and Collaboration in Interdisciplinary Research. SAGE, Thousand Oaks, CA.

Morin, A. 2011: Self-awareness part 2: Neuroanatomy and importance of inner speech. Social and Personality Psychology Compass 5:1004–1017.

O'Rourke, M., and S. J. Crowley. 2013: Philosophical intervention and cross-disciplinary science: The story of the Toolbox Project. Synthese 190:1937–1954.

O'Rourke, M., S. J. Crowley, and C. Gonnerman. 2016: On the nature of cross-disciplinary integration: A philosophical framework. Studies in History and Philosophy of Science Part C: Studies in History and Philosophy of Biological and Biomedical Sciences 56:62–70.

Okhuysen, G. A., and K. M. Eisenhardt. 2002: Integrating knowledge in groups: How formal interventions enable flexibility. Organization Science 13:370–386.

Parker, S. K., P. W. B. Atkins, and C. M. Axtell. 2008: Building better work places through individual perspective taking: A fresh look at a fundamental human process. Pp. 149–196 *in* G. Hodgkinson and K. Ford, eds. International Review of Industrial and Organizational Psychology. Vol. 23. Wiley, Chichester.

Quinlan, E., and S. Robertson. 2010: Mutual understanding in multi-disciplinary primary health care teams. Journal of Interprofessional Care 24:565–578.

Repko, A. F. 2012: Interdisciplinary Research: Process and Theory. SAGE, Thousand Oaks, CA.

Rinkus, M. A., S. Donovan, T. E. Hall, and M. O'Rourke. 2020: Using a survey to initiate and sustain group dialogue in focus groups. International Journal of Social Research Methodology. Forthcoming. DOI: 10.1080/13645579.2020.1786240.

Ruby, P., and J. Decety. 2001: Effect of subjective perspective taking during simulation of action: A PET investigation of agency. Nature Neuroscience 4:546–550.

Salazar, M. R., K. Widmer, K. Doiron, and T. K. Lant. 2019: Leader integrative capabilities: A catalyst for effective interdisciplinary teams. Pp. 313–328 *in* K. L. Hall, A. L. Vogel, and R. T. Croyle, eds. Strategies for Team Science Success: Handbook of Evidence-Based Principles for Cross-Disciplinary Science and Practical Lessons Learned from Health Researchers. Springer, New York.

Schiffer, S. R. 1972: Meaning. Oxford University Press, Oxford.

Settles, I. H., N. T. Buchanan, and K. Dotson. 2019: Scrutinized but not recognized:(In) visibility and hypervisibility experiences of faculty of color. Journal of Vocational Behavior 113:62–74.

Stalnaker, R. 1978: Assertion. Syntax and Semantics 9:315–332.

———. 2002: Common ground. Linguistics and Philosophy 25:701–721.

Ta, V. P., M. J. Babcock, and W. Ickes. 2017: Developing latent semantic similarity in initial, unstructured interactions: The words may be all you need. Journal of Language and Social Psychology 36:143–166.

Thompson, J. L. 2009: Building collective communication competence in interdisciplinary research teams. Journal of Applied Communication Research 37:278–297.

Tsoukas, H. 2009: A dialogical approach to the creation of new knowledge in organizations. Organization Science 20:941–957.

West, M. A. 1996: Reflexivity and work group effectiveness: A conceptual integration. Pp. 555–579 *in* M. A. West, ed. Handbook of Work Group Psychology. Wiley, Chichester.

Widmer, P. S., M. C. Schippers, and M. A. West. 2009: Recent developments in reflexivity research: A review. Psychology of Everyday Activity 2:2–11.

11 Future Directions for the Toolbox Dialogue Initiative

Sanford D. Eigenbrode
Stephanie E. Vasko
Anna Malavisi
Bethany K. Laursen
Michael O'Rourke

11.1 INTRODUCTION

The narrative history in Chapter 3 of this volume describes how the Toolbox Dialogue Initiative (TDI) has evolved from being analogue to being digital. This characterization of TDI is both metaphorical and literal. It metaphorically describes the fact that TDI has remained committed to doing what it does while embracing the need to engage in continual improvement, making changes on the basis of evidence we gather in running workshops. It also literally describes the fact that TDI has grown from using paper forms in workshops aimed at helping graduate students talk about disciplinary differences to using an array of digital technologies that provide assistance to a wide range of clients who want to use philosophy to improve their communicative and collaborative capacity. This chapter describes both the most important recent changes for TDI as well as a number of future directions. After discussing new technologies for TDI's "Digital Age," we consider TDI relationships with teams that go beyond single workshops and involve more longitudinal engagement, strategies for building on TDI insights, and new directions for TDI research.

11.2 NEW TECHNOLOGIES FOR TDI'S DIGITAL AGE

As TDI grows, so do the ways in which it operates and conducts research. It has evaluated and improved its internal technological readiness as it has become digital. For more on "technological readiness," see Hall et al. (2008) and Olson and Olson (2000). During the early years of TDI, researchers manually entered data from paper instruments into software like Microsoft Excel®. TDI researchers now use a variety of digital tools to design, facilitate, and analyze workshops. TDI often uses Qualtrics® software to collect pre- and post-workshop surveys and for additional activities before, during, and after workshops. These additional activities include communications surveys conducted periodically during the lifetime of a project (e.g., see Chapter 12), workshop registration prior to workshops, and surveys conducted during workshops.

TDI collects rating-response data during most workshops using a web-based version of the Toolbox instrument. Participants bring a Wi-Fi-enabled device to the session and during the session they are directed to a link for the workshop-specific instrument. Each participant is assigned a unique code in order to identify them in the collected data.

In many workshops, TDI facilitators record the audio from the workshop with external microphones and software (typically, AudioNote®) that supports the annotation of each speaking turn. These annotations note the participant who produced the speaking turn and a few of the words that were spoken. Previously, speaking turn annotations were produced manually by the workshop recorder using a form described in Looney et al. (2014). For transcribed recordings, this software-based annotation helps link speaking turns with specific participants. TDI uses independent transcription services to transcribe its audio recordings. TDI verifies the transcriptions and assigns participant numbers to the speaking turns.

TDI now offers virtual workshops in addition to its traditional in-person offerings. When participants and the facilitator are unable to be together in one place together, TDI uses online meeting software with video and audio capabilities (typically Zoom®). TDI has created a set of internal best practices for facilitating virtual meetings; these include such tips as requiring each participant to use an adjustable microphone with an on/off function to ensure good sound quality during the workshop and for the audio recording.

TDI has also adopted online tools so that its members can share resources, communicate about projects (including this book), coordinate meetings at conferences, and update one another both synchronously and asynchronously. During workshops, these tools are used for exchanges between the facilitator and recorder and to help the facilitator navigate emotionally charged or difficult workshops with support from the TDI team. During workshops conducted in parallel, they are used to coordinate responses to questions, divide workloads, and address consistently any issues that arise. TDI stores documentation, data, and resources on a cloud-based collaboration platform that meets the requirements of the Michigan State University (MSU) Institutional Review Board. TDI members also use collaborative tools from Google to work with each other and with non-TDI collaborators.

TDI has created a living data management plan to manage the large volume of data that has been collected over the years. It defines what files should include, how files should be named, where files should be stored based on their contents, acceptable file formats, what items from each workshop should be shared online, and how the log of workshops should be updated. The initial plan was created by a team that included MSU members of the TDI community using DMPTool (University of California Curation Center 2019). TDI now requires that one of its community members serve as the TDI data manager. One of their tasks is to periodically update the data management plan, making it flexible enough to adapt to changes in staff, tools, and research protocols. Developing a clear and dynamic data management strategy is especially important for TDI because grant agencies increasingly require data sharing, open data, and a data management plan.

As TDI continues to mature in its digital era, it is also exploring state-of-the-art technologies for analyzing Toolbox data. TDI has always sought to be a leader in interdisciplinary science and team collaboration and in this spirit it continues to explore and develop the latest technologies, such as machine learning. The goal is to provide deeper insights and understanding, especially to the field of the science of team science.

11.3   BEYOND SINGLE-SESSION INTERVENTIONS: LONGITUDINAL ENGAGEMENT WITH TEAMS

Toolbox workshops were conceived and designed to enhance communication among the members of collaborative teams by fostering deeper mutual understanding of team members' assumptions and commitments about science and its applications (Eigenbrode et al. 2007). An assumption, borne out through experience, was that a single, well-executed Toolbox workshop can have lasting, beneficial effects. A good workshop improves the awareness of individual team members, their capacity to articulate their own assumptions and views, and their ability to hear and understand others (see Chapters 5, 7, 9, 10). It also can influence a team's entire approach to working together, informing its culture of collaboration, through the experience of sharing fundamental views and assumptions in a structured dialogue. However, a single workshop has not been enough for some teams. On occasion, participants in a workshop have later expressed interest in reprising a Toolbox-style workshop in order to explore certain issues more deeply or to reexamine them as they arise during the project. Although the potential need for this was recognized early (Eigenbrode et al. 2007), TDI has only recently engaged longitudinally with teams.

There are three reasons for such longitudinal engagement. The first has to do with the life cycle of a project. Collaborative projects typically have identifiable phases, which include forming the team, building the team, developing methods, doing the science, reporting results, evaluating the process, output and outcomes, and taking care of the project's legacy (Lang et al. 2012; Eigenbrode et al. 2017). Each of these phases presents distinct philosophical challenges. For example, during team formation and building, it may be especially vital to address issues related to motivation and values. During development of methods, it can become important to address issues related to validation, reality, and ways of knowing. When communicating project findings to stakeholders, the public, and peers, issues concerning the roles of values and advocacy can be important. Although each of these philosophical challenges is addressed in a single Toolbox workshop, they often merit reexamination in one or more subsequent workshops.

The second reason motivating longitudinal engagement concerns *team turnover*. Many projects, especially those involving graduate students, experience personnel turnover. For example, NSF Integrative Graduate Education and Research Traineeship (IGERT) projects, which have been an important platform for developing TDI methods (see Chapter 3), often involve two or more cohorts of graduate students. (In some of these projects, a Toolbox workshop was part of cohort initiation.) At the University of Idaho (UI), IGERT projects on sustainable production and biodiversity conservation had two cohorts, each of which participated in a Toolbox workshop (Bosque-Pérez et al. 2016). A water resources-focused IGERT project at UI had three cohorts, each of which participated in a workshop. The Toolbox workshops improved communication within each cohort and also helped communication and integration among cohorts. Teams with planned or unplanned turnover can realize similar benefits from multiple Toolbox workshops. Many projects addressing complex issues affecting human and natural ecosystems must be sustained long enough that they will likely experience turnover in personnel. They typically need mechanisms for introducing new team members to the project and its collaborative culture. Delving into philosophical issues through Toolbox workshops can improve mutual understanding of philosophically based differences and can also establish bonds between new and veteran team members.

Finally, the presence of philosophical "hot spots" within a project team can keep the Toolbox approach relevant over time. Most Toolbox workshops include especially productive discussion around one or a few of the core questions, which may reveal cognitive and philosophical differences that can affect team performance. These hot spots may require subsequent investigation, whether via individual reflection, one-on-one conversation, or continued dialogue among participants. However, hot-spot issues may not be simple enough to resolve through heightened awareness or informal interactions. In such cases, it is often valuable for the team to explore these hot spots more deeply in one or more additional Toolbox workshops. For example, many teams struggle with the subtle issues related to potential bias, the presence and merits of advocacy, and the nature of proper relationships with stakeholders. These struggles can affect project design, research execution, and implementation of results. Another hot spot can be differences in views about the validity of qualitative and quantitative data and the appropriate methods for their integration. A Toolbox workshop discussion dedicated to these topics can illuminate them but may not resolve them well enough for project execution. To address such hot spots, a team may decide to undertake follow-up Toolbox workshops focused only on the most productive prompts or on new prompts designed for the purpose of addressing one issue (see *Targeted Workshops* below).

11.4   STRATEGIES FOR BUILDING ON TDI INSIGHTS

There are several strategies that can be used separately or together to address the challenges described in the preceding section and to build on an initial Toolbox workshop as a project continues. Their design and implementation will be project specific. We have employed most of these as part of longitudinal engagement with teams. Some of the projects that have helped us formulate these strategies include Idaho's "Genes by Environment: Modeling, Mechanisms, Mapping" team, or *GEM3* (NSF-EPSCoR, www.idahogem3.org), "Woody Invasive Alien Species in East Africa: Assessing and Mitigating their Negative Impact on Ecosystem Services and Rural Livelihood," or *Woody Weeds* (Swiss National Science Foundation, www.cabi.org/woodyweeds), three separate NSF-IGERT projects at UI, and "Regional Approaches to Climate Change for Pacific Northwest Agriculture," or *REACCH* (USDA-NIFA CAP, www.reacchpna.org). For further details about several of these projects see Chapter 12. For a summary of the strategies and example workshops in which they've been applied, see Table 11.1.

11.4.1   Post-Workshop Reflection and Integration

The post-workshop report generated by TDI highlights prompts that elicited vigorous discussion as well as the prompts that revealed strong differences of opinion, as evidenced by marked heterogeneity of rating-response scores among participants.

TABLE 11.1

A Summary of Strategies Used to Build upon TDI Insights, and Examples of Projects That Have Used These Strategies

Strategies for Building on TDI Insights	Examples
Post-workshop reflection and integration	GEM3, Columbia University nursing course
Targeted workshops	UI IGERT project
Blended workshops	C4I strategic planning
Participant-designed workshops and prompts	Columbia University nursing course
Progressive Toolbox workshops	REACCH; Woody Weeds; UI IGERT project
The TDI "Solera" method	Woody Weeds
Action-step workshops	Woody Weeds
New educational applications	"Values and Responsibility in Interdisciplinary Environmental Science" (http://eese.msu.edu/)
Stakeholder workshops	West Michigan Sustainable Business Forum; Woody Weeds
Follow-on co-creation activities	Woody Weeds; REACCH
Co-publishing	Bosque-Perez et al. 2016; Knowlton et al. 2014; Hessels et al. 2015; Schnapp et al. 2012; Read et al. 2016

In addition, the report includes recommendations for enhancing collaborative aspects of team performance. All of these can be a basis for the improvement of future project activities. The TDI report can be shared to raise awareness of various aspects of the workshop that the TDI facilitators thought were especially important, including differences that could prove to be problematic and similarities that can be used to the team's benefit. Going forward, team members can monitor how these issues affect their collaboration, either through targeted discussion in team meetings or with project-wide surveys. Team members can be encouraged to be attentive to the issues and to develop activities to address them. All of our longitudinal engagements have profited from ongoing reflection on the Toolbox workshops we've conducted.

11.4.2 Targeted Workshops

The members of a project team can use a prompt-structured dialogue involving a targeted set of core questions in order to address hotspots detected in an initial Toolbox workshop or new issues that arise as the project matures. This can be done with or without assistance from TDI. Targeted workshops of this sort have been part of longitudinal engagement by TDI with several teams (see Chapter 12). For example, members of one IGERT team requested a special workshop to probe issues around integrating economics into their project. With TDI support, they generated a set of prompts focused on this issue that the team then discussed. In this instance, the five students and their faculty mentors collaboratively produced the prompts and then gathered to discuss them. TDI encourages teams to pursue such follow-up activities and can provide advice and facilitation.

11.4.3   BLENDED WORKSHOPS

Many issues facing a team during its project are logistical, such as developing and adhering to data management protocols, allocating effort to data analysis, and assigning writing tasks for publications. Successful resolution of these issues requires clear understanding of expectations, which can be facilitated by dialogue. We have conducted "blended" workshops, which contain dialogue prompts related to these logistical details as well as other prompts that continue to probe the philosophical dimensions of the collaboration. The habits of mind and of communication that are exercised during Toolbox workshops can help the team address these logistical issues. Sometimes, they are manifestations of deeper issues that can be addressed with more overtly philosophical prompts. For example, concern about author order can indicate deeper difficulties related to power imbalances or value differences. Awareness of these connections can help resolve what seem to be "merely" logistical challenges. Blended Toolbox workshops were a key part of the strategic planning process conducted in 2018 by the MSU Center for Interdisciplinarity (O'Rourke et al. 2020). These workshops were designed to generate input from the MSU community about the mission and vision of the new center, and several prompts concerned logistical and institutional aspects of interdisciplinary practice at MSU.

11.4.4   PARTICIPANT-DESIGNED WORKSHOPS AND PROMPTS

As noted in Chapters 2 and 3, TDI is committed to working with workshop partners when designing the instruments used to structure workshop dialogue. This collaboration familiarizes the partners with the premise and typical form of a Toolbox instrument. As a result, they are enabled to design follow-up activities to address specific logistical or philosophical issues. Participant-designed questions and prompts are likely to elicit committed engagement and productive discussion by team members. The development and exploration of such "homegrown" prompts enhances a project's collaborative culture and effectiveness.

TDI has had several partners engage in follow-up activities based on the Toolbox dialogue method. For example, Hessels et al. (2015) describe a semester-long implementation of the Toolbox approach in a course at Columbia University for nurses and pre- and post-doctoral fellows. It began with a Toolbox workshop structured by the Health Sciences Toolbox instrument (Schnapp et al. 2012). This workshop took place early in the semester, after which students in the course were divided into interdisciplinary groups of four or five. Over the course of the semester, there were three phases of engagement with the topics broached by the instrument. In the first, each group discussed three modules from the Health Sciences Toolbox instrument and added original prompts to each module. These new prompts expanded the module so that it more directly engaged with their specialties. In the second phase, each group received a group profile prepared by the course leaders that was based on the initial Toolbox workshop and on the work done in the first phase. This profile identified prompts that each group was advised to use in order to structure further collective reflection. In the third phase, each group worked through the remaining three modules and then evaluated the modules in a course meeting toward the end of the semester. This served to enhance each group's emerging interdisciplinary identity.

11.4.5 PROGRESSIVE TOOLBOX WORKSHOPS

For some teams, it can be useful to formally revisit TDI themes during a project. One way to do so is to invite project members to revise the instrument used in their initial Toolbox workshop. For example, students and postdocs in the REACCH project, a large agricultural and climate change project, created a project-specific instrument two years after an initial STEM-focused workshop. This instrument reflected its commitment to transdisciplinary research involving contributions from academics and non-academics (see Chapter 1). The project-specific instrument that was created was built around six core questions:

- *What are the requisites for successful transdisciplinary research?*
- *What motivates participants in transdisciplinary research?*
- *What are the hallmarks of successful, large transdisciplinary research projects?*
- *Must institutions be changed to adequately address complex issues like climate change?*
- *What are the benefits of sharing data in transdisciplinary research?*
- *How should researchers engage with stakeholders in transdisciplinary projects?*

The intent of these questions was to reveal attitudes about the critical dimensions and challenges of the project. Participants received guidance from TDI consultants, and the instrument was discussed in an annual project meeting that involved over 100 people. Individual prompts ranged from the more logistical, e.g., "*Graduate students should be strongly encouraged to collaborate with students from other disciplines,*" to the more philosophical, e.g., "*Transdisciplinary research is more difficult than disciplinary research*" (see Figure 11.1). Team member reactions to this exercise were divided. Some felt that the prompts encouraged conversation about and awareness of challenging issues. Others felt that it was important to return to the fundamental questions about interdisciplinary communication among the participating academic researchers.

A similar approach was used for the Woody Weeds project (see Chapter 12). TDI personnel first conducted a workshop with the project team during the launch meeting as a baseline, using the Scientific Research Toolbox instrument (see Appendix A). This instrument was then modified by TDI consultants and project team members and used in a second Toolbox workshop that took place approximately seven months after the initial one. Just a few prompts were added. For example, the prompt "*Sampling designs (e.g., number of samples, sampling area extent, and selection of indicators) must work for all spatial scales*" was added to the Methodology module in order to address a pressing challenge faced by the team. The prompt "*Environmental data represent reality and socio-economic data represent perception*" was added to the Reality module after being suggested during the debriefing session after the launch meeting of the project. This prompt reflected a prevailing attitude of some in the project and a willingness to explore controversy.

Motivation

Core Question: What motivates participants in transdisciplinary research?

1. Transdisciplinary research is as valuable/important as disciplinary research.

   Disagree   *Agree*

   1   2   3   4   5   *I don't know*   *N/A*

2. The potential of transdisciplinary research to address complex problems is the primary incentive for participation.

   Disagree   *Agree*

   1   2   3   4   5   *I don't know*   *N/A*

3. Individual motivation for transdisciplinary research originates from funding requirements.

   Disagree   *Agree*

   1   2   3   4   5   *I don't know*   *N/A*

4. The additional time and organizational requirements of participating in transdisciplinary research negate its benefits.

   Disagree   *Agree*

   1   2   3   4   5   *I don't know*   *N/A*

FIGURE 11.1 The Motivation module from the Regional Approaches to Climate Change (REACCH) Toolbox instrument.

If a project has sequential cohorts, each can undertake a Toolbox workshop using prompts added to the instrument by the preceding cohort. This worked well for the aforementioned IGERT project at UI that focused on water resources issues. The first cohort used the standard Scientific Research Toolbox instrument, to which they added prompts for the second cohort such as "Methodology: '*All members of a collaborative team must identify and agree upon common hypotheses*' and '*Teams function best with a designated leader to steer the direction of the team*' ." The issues that had challenged the first cohort are evident in the wording of these prompts.

11.4.6 THE TDI "SOLERA" METHOD

When modifying Toolbox instruments during an ongoing project, we find it useful to consider what we call the "solera" sequence of workshops. This means that the original instrument is revised to reflect emerging concerns within a project even as certain prompts are retained. ("Solera" is a Spanish word for a method of producing a blended wine whose taste reflects the multiple constituent wines.) This allows assessment of changes in the project over time and of differences in responses by incoming cohorts that could influence project functioning. When teams conduct a series of workshops whose instruments are developed in this manner, all but the first workshop address a combination of new and old topics. This helps teams face new issues without leaving behind long-standing challenges that may require continued engagement. This method was a key part of the development of the second Toolbox workshop delivered for the Woody Weeds project.

11.4.7   ACTION-STEP WORKSHOPS

Although structured dialogue improves team communication about key philo-sophically related issues, the goal of TDI is to improve overall communication and general team functioning. Initial Toolbox workshops sometimes suggest specific steps that teams can take to address communication difficulties or to deliberately incorporate diverse perspectives into team outputs. Consider again the example of the Woody Weeds project. Following two standard workshops in the project's first year, the workshop design was modified in the second year to focus on just four themes:

- *Ways of knowing, specifically, the relationship between scientific knowledge and indigenous knowledge*
- *Methodological difference, specifically, differences between qualitative and quan-titative reasoning*
- *Effective communication, specifically, transparency in relationships within the project*
- *Student perspectives, specifically, how the PhD students are integrated into the project*

Each of these themes was examined and elaborated on in breakout sessions. For example, when the first theme was considered, the prompt selected was "Indigenous knowledge and scientific knowledge are legitimate but irreconcilable." After exploring this issue deeply, the group prepared a set of responses that the project could use when the two sorts of knowledge diverged and needed resolution. Similar steps were developed by the other breakout groups. This process was helpful because the group had prior experience exploring these difficult issues due to the protection afforded by using the TDI method. Through this work, the group had learned that these sorts of issues stem from deeply held beliefs that require practice in reflection and sharing.

11.4.8   NEW EDUCATIONAL APPLICATIONS

Educational applications are a central part of the Toolbox identity. The approach grew out of a project intended to facilitate graduate education (Bosque-Pérez et al. 2016) and has been continually developed for graduate and undergraduate educa-tion (Knowlton et al. 2014; Hessels et al. 2015; Kjellberg et al. 2018). A prominent example of this commitment is the recent US NSF-sponsored Ethics Education in Science and Engineering project, "Values and Responsibility in Interdisciplinary Environmental Science" (http://eese.msu.edu/), which was built around the Toolbox approach. This effort involved both education and research components: the pro-ject team developed and implemented a dialogue-based curriculum that involved the Toolbox approach, and they also designed a systematic evaluation to test its effectiveness (Hall et al. 2018). The curriculum focuses on "social ethics," or "the collective social responsibility of the profession" (Herkert 2005, p. 374). The key Toolbox innovation here was the development of a curriculum that guides graduate students in analyzing the philosophical dimensions of their interdisciplinary space and in writing Toolbox-style prompts that structure dialogues conducted as part of the course implementation. The analysis and dialogue put the participants in close contact with important ethical assumptions that guide their practice as interdisciplinary environmental scientists.

11.4.9 STAKEHOLDER WORKSHOPS

Cross-disciplinary research often requires engagement with non-academic stakeholders from project conception through execution (Tress et al. 2004; Klein 2017). There are philosophical dimensions of practice that relate to the validation of knowledge claims and to the place of objectivity or the need to adhere to ethical norms. These dimensions cross academic and non-academic boundaries and are epistemological and metaphysical. Academic and non-academic stakeholders can have different assumptions, values, and ways of knowing that arise from disparate epistemological and metaphysical stances. TDI's ability to engage multiple disciplines and stakeholders can help to reveal such heterogeneity.

An early example of a stakeholder workshop focused on climate adaptation in West Michigan. Working with the West Michigan Sustainable Business Forum (WMSBF; https://wmsbf.org/), TDI contributed to the development of a framework for structuring region-scale climate adaptation efforts. These efforts, organized under the West Michigan Climate Resiliency Framework Initiative, brought together stakeholders from a range of business and professional sectors who were interested in developing economically sustainable responses to extreme weather events brought on by climate change. TDI and WMSBF partnered with the RAND Corporation and the Sustainable Climate Risk Management Project to use values-informed mental modeling to identify central issues of concern for the stakeholders and to develop a Toolbox instrument that focused on these concerns (Mayer et al. 2017). This instrument was used to structure dialogue in 15 parallel Toolbox workshops at a Climate Resiliency Summit in October 2014. These workshops have had a positive effect on the Climate Resiliency effort, which is still ongoing with participation by WMSBF and other partners.

A second example occurred as part of the Woody Weeds project. We conducted a Toolbox-inspired workshop for non-academic stakeholders that included pastoralists, village leaders, representatives of non-governmental organizations (NGOs), and government representatives from the Afar Region of Ethiopia. This mix of stakeholders made integration of Indigenous knowledge and scientific knowledge desirable if not necessary. There can be challenges and opportunities when trying to do this (Bohensky and Maru 2011). For example, concepts akin to replication and triangulation are often used for validation by non-scientists and scientists. However, non-scientists using Indigenous knowledge can also rely on authority and adherence to tradition, where the authority resides in preceding generations of practice. Whether these approaches can coexist, let alone inform or reinforce one another, depends again on effective communication and respectful processes of sharing knowledge (e.g., Gratani et al. 2011). These can be developed by using the Toolbox approach.

For example, a Validation module can include prompts allowing for responses that privilege tradition over certain kinds of observation. Some core questions that can help delineate potentially discordant positions are:

- *When conflicts arise from responses to environmental challenges, how should they be resolved?*
- *What is the proper relationship between human societies and ecosystems?*
- *Where should authority lie when deciding on responses to environmental challenges?*

Prompts for, say, the second question could be:

- *The natural order involves utilizing environmental resources to maximize human well-being.*
- *Humans should strive to alter ecosystems as little as possible.*
- *An environmentalist perspective is a luxury that does not make sense in our region.*
- *It is possible to find a balance between extraction of resources and preservation of ecosystem services.*
- *Since resources are limited, ultimately humans must compete for them.*

Facilitated discussions about this kind of fundamental issue can establish conditions for communication and clarification of differences and reveal the challenges and opportunities for effective collaboration.

The Woody Weeds workshop employed questions and prompts like those above but commenced with a core question and prompts designed to complement a prior "Local Implementation Group" workshop that was designed to facilitate the implementation of a management plan for woody invasive plants. A core question in the Toolbox workshop was: *Was the Local Implementation Group process completed last month successful?* Example prompts included: *The process allowed all participants to provide input equally*, and *I now better understand other stakeholders' positions.* Core questions and prompts were posted electronically and participants indicated their level of agreement using a hand-held remote. The results were immediately shared with the group.

11.4.10 FOLLOW-ON CO-CREATION ACTIVITIES

For multi-year interventions, many other models for integrating TDI insights into the culture of communication can be devised. Project leaders that are deliberate about enhancing collaborative skills and effectiveness will want to draw upon the resources and tools available for this purpose. These include collaborative software that help to develop decision-making and prioritizing processes (see Team Science Toolkit, www.teamsciencetoolkit.cancer.gov/). Teams addressing their collaboration challenges with a Toolbox workshop can use it as an introduction to these tools. Themes identified in an initial workshop can be the basis for subsequent actions, changes in practice, and interventions. TDI consultants can guide this process. For example, as the Woody Weeds project has progressed, issues that have emerged through Toolbox workshops have been addressed in focused follow-up workshops. The workshop that focused on four persistent communication issues (see "Action-Step Workshops" above) is an example of a follow-up model that can motivate co-creation activities. These include the creation of action steps the project can pursue in order to address the issues. In another example, Toolbox-based discussions in REACCH highlighted challenges with data protocols, and a project-wide "change team" was formed. It worked to improve understanding and compliance with data management protocols. The difficulty faced by the project was largely one of communication, but there were also differences in conviction and expectation for data ownership, rights, and responsibilities by individual researchers.

11.4.11 CO-PUBLISHING

TDI is dedicated to the investigation of how communication works within teams and the development of best practices for promoting better communication (see Chapter 5). TDI views our clients as partners in discovery. Longitudinal engagement with some teams is integral to their efforts to pioneer new approaches to integration. For example, TDI engagement with a University of Idaho IGERT project was described in an article that presented the project's approach to team-based graduate education (Bosque-Pérez et al. 2016). Additional articles written with workshop partners concern environmental education (Knowlton et al. 2014), nursing education (Hessels et al. 2015), translational medicine (Schnapp et al. 2012), and team science (Read et al. 2016).

11.5   NEW DIRECTIONS IN RESEARCH

As noted in Chapters 2, 3, 9, and 10, TDI is dedicated to research as well as to facilitation. Just as it has expanded the scope of its facilitation and increased the size of its community, TDI has also increased and diversified its research portfolio. Chapter 2 describes the research of TDI as having three facets: providing evidence-based feedback to the partner teams, assessing the impact of the Toolbox approach, and characterizing the nature of cross-disciplinary process. In what follows, we focus on new research directions that fall into the last two categories.

11.5.1 ASSESSMENT OF TOOLBOX IMPACT

Part of our assessment of Toolbox impact involves the investigation of new contexts in which the approach might prove useful. For example, TDI has explored the potential for Toolbox dialogue to enhance communication for teams engaged in global development and humanitarian intervention. Such teams confront many challenges. First, there can be difficulties communicating across organizations and communities. Concepts such as *development* and *sustainability* may have different meanings among team members. Communication challenges can be exacerbated by differences among the natural languages spoken by professionals and the communities involved, as well as by differences in the technical languages they speak. Second, it can be challenging to assess whether development actions generate more harm than good for some members of the target communities. What assumptions are made, and by whom, about the implications of development actions for gender equity in these communities? Although in some cases aid has ameliorated the situation of women, it has worsened it in other cases (Harding and Malavisi 2017). Third, injustice and inequality in global development are partially attributable to epistemic domination and marginalization, which harm populations that are denied the status of knowers by those with power and privilege (Dotson 2012). This can take the form of *credibility deficit*, which can involve attribution of less credibility to the knowledge of, for example, Indigenous people by local and international NGOs.

TDI has begun to facilitate workshops in which these communicative, ethical, and epistemic challenges are highlighted for aid professionals. Such practitioners may rarely or never reflect on these important challenges. Inconsistencies among worldviews, values, and assumptions are often not discussed. There is a need to increase awareness of these inconsistencies through reflexive processes such as Toolbox workshops, which can enable collaborators to engage in a dialogue about tacit assumptions that underlie their worldviews. Toolbox workshops can also support analyses of assumptions, power dynamics, biases, ethical issues, and epistemic injustice within organizations and communities working on sustainable development and humanitarian intervention. Structured dialogue can render these issues more salient and give team members an opportunity to confront and discuss them openly.

In structuring these dialogues, we use a feminist methodology that is grounded in non-ideal theory, i.e., theory that recognizes the "centrality of oppression" and the importance of situational location (Mills 2005, p. 174). Although grounded in theory, this methodology has important implications for practice, e.g., by directing us to structure dialogue among development professionals that calls attention to injustices such as the persistent oppression of minority groups (Parekh and Wilcox 2018). This methodology also demands particular commitments that reflect fundamental aspects of feminist theory, such as intersectionality, context sensitivity, and the development of self-reflexive critiques (Harding 1987; Parekh and Wilcox 2018). These commitments also have implications for practice since they are considered to be an essential component of working with groups within the global development and humanitarian aid sector.

TDI has conducted three workshops with development professionals. One involved a global NGO, one involved an international conference on sustainable development attended by researchers and practitioners, and the third involved the annual InterAction conference, the largest network of NGOs in the United States. The modules used for these workshops were based on concepts central to the practice of aid: *development, justice, knowledge/cultural understanding*, and *sustainability*. In all three workshops, dialogue was engaged and thoughtful, which underscores the centrality of these concepts to the working lives and worldviews of the participants. This illustrates that Toolbox workshops have value even when the communicative, ethical, and epistemic challenges are very complex and difficult because the groups involved are composed of many different types of stakeholders.

11.5.2   CHARACTERIZING CROSS-DISCIPLINARY PROCESS

Our interest in cross-disciplinary process involves a number of philosophical issues. These include disagreement (Crowley et al. 2016) and ignorance (Piso et al. 2016).

We have also conducted research on critical aspects of cross-disciplinary research and practice. These include communication (Hall and O'Rourke 2014) and the role of values (Robinson et al. 2016). One topic, *integration*, is of great interest to us because it is philosophical as well as critical to cross-disciplinary research and practice. The combination of different perspectives is an essential part of cross-disciplinary success. TDI has created a model of cross-disciplinary integration known as the "Input-Process-Output" (IPO) model (O'Rourke et al. 2016), which we describe below. We apply and extend that theory by examining a range of integrative phenomena (e.g., O'Rourke et al. 2019). This theory-based approach to integration aids our investigation of the integrative aspects of Toolbox dialogue (see Chapters 5 and 7), which can yield valuable insights for the teams that work with us. These applications of the theory also present us with new data that need to be explained and thereby help us refine the theory.

Members of TDI have created an integration working group, which engages in theoretical projects as well as in projects intended to make sense out of Toolbox data. This group is developing an epistemic taxonomy of integrative episodes based on the inputs, process, and outputs of integration. Initial work on this is reported in Chapter 5. The group is also identifying and modeling what we call "micro-integrative moments" in Toolbox dialogue. These are compelling instances of integrative relations among inputs. This modeling effort highlights the operation of integrative processes at finer scales of interdisciplinary activity. A key integrative process at these scales is collaborative, interdisciplinary reasoning, which uses group argumentation to secure integrative relations (Laursen 2018; Laursen and O'Rourke 2019). In a third project, the group is using virtue epistemology to elaborate the claim that integration is an epistemic virtue of teams that is determined by their structure, dynamics, and composition. It is a trait of teams that successfully combine their perspectives in pursuit of a research response that is commensurable with the problem being addressed. Finally, the group is continuing to develop a taxonomy of integrative relations, which builds on O'Rourke et al. (2016). The goals of this last project are to develop a new discourse analysis method and to elaborate the possible integrative relations one can find in cross-disciplinary activities.

All of these projects build on the IPO model of integration of O'Rourke et al. (2016). This model is comprehensive and flexible, but it is not specific enough to provide guidance for those interested in observing integration in practice. The aforementioned projects are efforts by TDI to elaborate the schematic IPO model and render it useful for other cross-disciplinary theorists and practitioners.

TDI has undertaken a long-term effort to improve how cross-disciplinary integration is measured and evaluated, which directly supports our mission to conduct engaged philosophy that improves practice. The initial phase of this collaborative effort with colleagues at the National Socio-Environmental Synthesis Center (SESYNC) is a systematic review of published attempts to assess interdisciplinarity (Laursen et al. 2020). We find that philosophical assumptions about why assessment is important, how interdisciplinary processes work, and who can contribute to assessment are holding the field back. These assumptions are especially pronounced when it comes to integration, rendering it the most difficult aspect of interdisciplinarity to assess. These findings are consistent with previous findings (e.g., Wagner et al. 2011) and they show how these philosophical assumptions manifest and influence specific assessment practices such as bibliometric indexes (Bordons et al. 2004) and critical questions (Späth 2008). Given these results, TDI is using insights from our various integration projects to develop a replicable and transparent methodology for assessing cross-disciplinary integration. The methodology focuses on integrative relations, which are a key part of the analysis of integrative process in the IPO model. Our goal is to develop an assessment methodology that reliably tracks cross-disciplinary integration and can be implemented without expert knowledge. We intend to combine the unique features of the IPO model and principles of measurement and evaluation so as to develop applications of the model that help us accomplish three goals: inform deliberation among collaborators in cross-disciplinary efforts, evaluate their cross-disciplinary efforts, and better understand cross-disciplinary processes.

11.6 ACKNOWLEDGMENTS

We are grateful to our colleagues and the Toolbox workshop participants for their contributions to the new directions described in this chapter. We are especially grateful to Russell Werner for designing and maintaining the Toolbox app, and to Marisa A. Rinkus for her help in developing the data management plan. We thank Graham Hubbs and Steven Orzack for their editing assistance. O'Rourke's work on this chapter was supported by the USDA National Institute of Food and Agriculture, Hatch Project 1016959.

11.7 LITERATURE CITED

Bohensky, E. L., and Y. Maru. 2011: Indigenous knowledge, science, and resilience: What have we learned from a decade of international literature on "integration?" Ecology and Society 16:1–20.

Bordons, M., F. Morillo, and I. Gómez. 2004: Analysis of cross-disciplinary research through bibliometric tools. Pp. 437–456 in H. F. Moed, W. Glänzel, and U. Schmoch, eds. Handbook of Quantitative Science and Technology Research. Kluwer, New York.

Bosque-Pérez, N. A., P. Z. Klos, J. E. Force, L. P. Waits, K. Cleary, P. Rhoades, S. M. Galbraith, A. L. B. Brymer, M. O'Rourke, S. D. Eigenbrode, B. Finegan, J. D. Wulfhorst, N. Sibelet, and J. D. Holbrook. 2016: A pedagogical model for team-based, problem-focused interdisciplinary doctoral education. BioScience 66:477–488.

Crowley, S. J., C. Gonnerman, and M. O'Rourke. 2016: Cross-disciplinary research as a platform for philosophical research. Journal of the American Philosophical Association 2:344–363.

Dotson, K. 2012: A cautionary tale: On limiting epistemic oppression. Frontiers: A Journal of Women Studies 33:24–47.

Eigenbrode, S. D., M. O'Rourke, J. D. Wulfhorst, D. M. Althoff, C. S. Goldberg, K. Merrill, W. Morse, M. Nielsen-Pincus, J. Stephens, L. Winowiecki, and N. A. Bosque-Pérez. 2007: Employing philosophical dialogue in collaborative science. BioScience 57:55–64.

Eigenbrode, S. D., T. Martin, L. W. Morton, J. Colletti, P. Goodwin, R. Gustafson, D. Hawthorne, A. Johnson, J. T. Klein, L. Mercado, S. Pearl, T. Richard, and M. Wolcott. 2017: *Leading large transdisciplinary projects addressing social-ecological systems: A primer for project directors*. Downloaded from https://nifa.usda.gov/leading-transdisciplinary-projects.

Gratani, M., J. Butler, F. Royee, P. Valentine, D. Burrows, and A. Anderson. 2011: Is validation of indigenous ecological knowledge a disrespectful process? A case study of traditional fishing poisons and invasive fish management from the Wet Tropics, Australia. Ecology and Society 16:25.

Hall, K. L., D. Stokols, R. P. Moser, B. K. Taylor, M. D. Thornquist, L. C. Nebeling, C. C. Ehret, M. J. Barnett, A. McTiernan, N. A. Berger, M. I. Goran, and R. W. Jeffery. 2008: The collaboration readiness of transdisciplinary research teams and centers: Findings from the National Cancer Institute's TREC year-one evaluation study. American Journal of Preventive Medicine 35:S161–S172.

Hall, T. E., and M. O'Rourke. 2014: Responding to communication challenges in transdisciplinary sustainability science. Pp. 119–139 *in* K. Huutoniemi and P. Tapio, eds. Heuristics for Transdisciplinary Sustainability Studies: Solution-Oriented Approaches to Complex Problems. Routledge, Oxford.

Hall, T. E., Z. Piso, J. Engebretson, and M. O'Rourke. 2018: Evaluating a dialogue-based approach to teaching about values and policy in graduate transdisciplinary environmental science programs. PLOS ONE 13:e0202948.

Harding, S. 1987: Feminism and Methodology. Indiana University Press, Bloomington.

Harding, S., and A. Malavisi. 2017: Women, gender, and philosophies of global development. Pp. 419–431 *in* A. Garry, S. Khader, and A. Stone, eds. The Routledge Companion to Feminist Philosophy. Routledge, New York.

Herkert, J. R. 2005: Ways of thinking about and teaching ethical problem solving: Microethics and macroethics in engineering. Science and Engineering Ethics 11:373–385.

Hessels, A. J., B. Robinson, M. O'Rourke, M. D. Begg, and E. L. Larson. 2015: Building interdisciplinary research models through interactive education. Clinical and Translational Science 8:793–799.

Kjellberg, P., M. O'Rourke, and D. O'Connor-Gomez. 2018: Interdisciplinarity and the undisciplined student: Lessons from the Whittier Scholars Program. Issues in Interdisciplinary Studies 36:34–65.

Klein, J. T. 2017: Typologies of interdisciplinarity: The boundary work of definition. Pp. 21–34 *in* R. Frodeman, J. T. Klein, and R. C. S. Pacheco, eds. The Oxford Handbook of Interdisciplinarity. Oxford University Press, Oxford.

Knowlton, J., K. Halvorsen, R. Handler, and M. O'Rourke. 2014: Teaching interdisciplinary sustainability science teamwork skills to graduate students using in-person and web-based interactions. Sustainability 6:9428–9440.

Lang, D. J., A. Wiek, M. Bergmann, M. Stauffacher, P. Martens, P. Moll, M. Swilling, and C. J. Thomas. 2012: Transdisciplinary research in sustainability science: practice, principles, and challenges. Sustainability Science 7:25–43.

Laursen, B. K. 2018: What is collaborative, interdisciplinary reasoning? The heart of interdisciplinary team science. Informing Science 21:75–106.

Laursen, B. K., and M. O'Rourke. 2019: Thinking with Klein about integration. Issues in Interdisciplinary Studies 37:33–61.

Laursen, B. K., N. Motzer, and K. Anderson. 2020: Results of a systematic review show how to improve measurement of interdisciplinarity. Presented at the 11th Annual Science of Team Science Conference, Virtual. https://youtu.be/q89CcZyHfr8.

Looney, C., S. Donovan, M. O'Rourke, S. J. Crowley, S. D. Eigenbrode, L. Rotschy, N. A. Bosque-Pérez, and J. D. Wulfhorst. 2014: Seeing through the eyes of collaborators: Using Toolbox workshops to enhance cross-disciplinary communication. Pp. 220–243 *in* M. O'Rourke, S. J. Crowley, S. D. Eigenbrode, and J. D. Wulfhorst, eds. Enhancing Communication and Collaboration in Interdisciplinary Research. SAGE, Thousand Oaks, CA.

Mayer, L. A., K. Loa, B. Cwik, N. Tuana, K. Keller, C. Gonnerman, A. M. Parker, and R. J. Lempert. 2017: Understanding scientists' computational modeling decisions about climate risk management strategies using values-informed mental models. Global Environmental Change 42:107–116.

Mills, C. W. 2005: "Ideal theory" as ideology. Hypatia 20:165–184.

O'Rourke, M., S. J. Crowley, and C. Gonnerman. 2016: On the nature of cross-disciplinary integration: A philosophical framework. Studies in History and Philosophy of Science Part C: Studies in History and Philosophy of Biological and Biomedical Sciences 56:62–70.

O'Rourke, M., S. J. Crowley, B. K. Laursen, B. Robinson, and S. E. Vasko. 2019: Disciplinary diversity in teams, integrative approaches from unidisciplinarity to transdisciplinarity. Pp.21–46 *in* K. A. Kall, A. L. Vogel, and R. T. Croyle, eds. Advancing Social and Behavioral Health Research through Cross-Disciplinary Team Science: Principles for Success. Springer, Berlin.

O'Rourke, M., S. E. Vasko, C. McLeskey, and M. Rinkus. 2020: Philosophical dialogue as field philosophy. Pp. 48–65 *in* E. Brister and R. Frodeman, eds. Using Philosophy to Change the World: A Guide to Field Philosophy. Routledge, New York.

Olson, G. M., and J. S. Olson. 2000: Distance matters. Human-Computer Interaction 15:139–178.

Parekh, S., and S. Wilcox. 2018: *Feminist Perspectives on Globalization.* Downloaded from https://plato.stanford.edu/archives/spr2018/entries/feminism-globalization/.

Piso, Z., E. Sertler, A. Malavisi, K. Marable, E. Jensen, C. Gonnerman, and M. O'Rourke. 2016: The production and reinforcement of ignorance in collaborative interdisciplinary research. Social Epistemology 30:643–664.

Read, E. K., M. O'Rourke, G. S. Hong, P. C. Hanson, L. A. Winslow, S. J. Crowley, C. A. Brewer, and K. C. Weathers. 2016: Building the team for team science. Ecosphere 7:e01291.

Robinson, B., S. E. Vasko, C. Gonnerman, M. Christen, M. O'Rourke, and D. Steel. 2016: Human values and the value of humanities in interdisciplinary research. Cogent Arts & Humanities 3:1123080.

Schnapp, L. M., L. Rotschy, T. E. Hall, S. J. Crowley, and M. O'Rourke. 2012: How to talk to strangers: Facilitating knowledge sharing within translational health teams with the Toolbox dialogue method. Translational Behavioral Medicine 2:469–479.

Späth, P. 2008: Learning ex-post: Towards a simple method and set of questions for the self-evaluation of transdisciplinary research. GAIA-Ecological Perspectives for Science and Society 17:224–232.

Tress, G., B. Tress, and G. Fry. 2004: Clarifying integrative research concepts in landscape ecology. Landscape Ecology 20:479–493.

University of California Curation Center. 2019: *DMP Tool.* Downloaded from https://dmptool.org/.

Wagner, C. S., J. D. Roessner, K. Bobb, J. T. Klein, K. W. Boyack, J. Keyton, I. Rafols, and K. Börner. 2011: Approaches to understanding and measuring interdisciplinary scientific research (IDR): A review of the literature. Journal of Informetrics 5:14–26.

12 Toolbox Workshop Case Studies

Graham Hubbs
Michael O'Rourke
Sanford D. Eigenbrode
Marisa A. Rinkus
Anna Malavisi

12.1 INTRODUCTION

This chapter presents eight case studies of Toolbox workshops. Our goal is to give the reader a sense of the variety of Toolbox workshop experiences. In these case studies, we discuss various groups TDI has worked with, the different ways in which we've structured the Toolbox workshop experience, and the range of activities we've used. This chapter may be of particular interest to the reader who is considering using the Toolbox approach, as it provides a sense of its flexibility and scalability.

Each case study is organized in similar fashion. We begin by describing the objectives of the workshop, both for our partner and for TDI. After describing the participants who took part in the workshop, we then discuss the instrument design process, illustrate some of the prompts used in the workshop, and detail the processes involved in setting the stage for each workshop and in the workshop itself. We conclude with a description of some of the outcomes of each workshop. For one workshop conducted with a long-term partner, we discuss the context of the workshop within that larger partnership.

We have organized these case studies chronologically, beginning with one that took place in 2010. This case study describes workshops that were run for a newly formed cohort of graduate students who had just joined a team-based Integrative Graduate Education and Research Traineeship (IGERT) project at the University of Idaho. It highlights what TDI can do for students who are just beginning their graduate work, particularly in cross-disciplinary, team-based settings.

The second case study, which took place in 2012, also addresses graduate student pedagogy. The workshops described in this case study were delivered at a retreat for graduate students and post-doctoral research fellows who were part of a large, multi-institutional agricultural project that focused on climate change impacts on US Pacific Northwest agriculture. This workshop used an instrument designed for projects that focused on climate change. In addition to presenting some common reactions participants have to the philosophical self-reflection encouraged by our workshops, this case study illustrates the impact Toolbox dialogue can have on the culture of a larger project.

The third case study describes our first collaboration with a non-academic partner. It involves 13 workshops held as part of the Climate Resiliency Summit in 2014 in Grand Rapids, Michigan. This case study highlights the scalability of the Toolbox approach, since these workshops were conducted in parallel and involved the participation of more than 80 local residents. They were the result of a partnership with a local business organization that enlisted us to help facilitate a region-wide effort to make West Michigan more resilient and sustainable in the face of climate change.

The fourth case study describes our work with an international project known as Woody Weeds, which focuses on invasive plants in East Africa. We discuss a workshop held in 2015, which was the second of several workshops we have conducted for this group. We include a synopsis of our long-term collaboration with this Swiss-based effort, which was described in Chapter 11. This case study illustrates the potential for the Toolbox approach to support complex, cross-disciplinary, multi-institutional, international, and transdisciplinary collaborations.

The workshop in the fifth case study was also designed for and delivered to non-academic participants. It was developed for the community of Winnebago County, Illinois, and held in 2016. TDI worked with a graduate student at the University of Idaho to design a novel Toolbox instrument that was intended to help the county residents reflect on their shared agricultural landscape in a time of social and political disagreement. It illustrates what the Toolbox approach can do to help members of smaller communities improve their collaboration and communication.

The workshop discussed in the sixth case study was run for Michigan State University's Office of Sustainability in 2016. The Toolbox instrument and workshop protocol were designed in collaboration with the Office of Sustainability. Workshop goals included fostering university-wide conversation about sustainability planning and gathering information about the values and priorities related to sustainability found on campus. This case study provides another example of the work we can do for non-research organizations, this time in a large institutional context.

The seventh case study describes the first workshop conducted for a development-focused non-governmental organization (NGO). In 2015, TDI began adapting the Toolbox approach for the global development and humanitarian aid sector, and this workshop, conducted in October 2016, marks the first time we implemented this new adaptation. After generating a new development-focused Toolbox instrument, we met with representatives of this large international NGO at their home office. This case study demonstrates the adaptability of the Toolbox approach to new, non-research contexts.

The workshops described in the final case study were conducted for several research teams funded by a large, cross-disciplinary initiative dedicated to tackling grand challenges. This foundation-sponsored initiative, which was organized by two centers at Purdue University, provided these research teams with a series of capacity-building opportunities near the beginning of their lifecycles. These Toolbox workshops occurred in 2017. This case study shows the benefit of TDI involvement at the beginning of large-scale, cross-disciplinary projects, particularly those seeking to integrate the natural sciences, social sciences, and humanities.

As noted in Chapter 11, we have also published detailed descriptions of a variety of Toolbox workshops (Schnapp et al. 2012; Knowlton et al. 2014; Hessels et al. 2015;

Hall et al. 2018; Kjellberg et al. 2018; Berling et al. 2019). Readers interested in more detailed accounts of Toolbox workshops that include evaluation of the approach are encouraged to consult these articles.

12.2 UNIVERSITY OF IDAHO IGERT GRADUATE STUDENT RESEARCH TEAM WORKSHOPS

This case study describes a set of Toolbox workshops conducted for graduate student research teams from the University of Idaho (UI) during 2010. Members of these teams were part of a new cohort of graduate students who had just matriculated into a team-based, graduate education program. These workshops highlight the value of this kind of capacity-building experience for graduate students who are at the beginning of their research training, especially graduate students in cross-disciplinary programs.

12.2.1 OBJECTIVES

Three workshops were held as part of a one-day immersive training for graduate students working on an IGERT project called "Evaluating Resilience of Ecological and Social Systems in Changing Landscapes: A Doctoral Research and Education Program in Idaho and Costa Rica" (Bosque-Pérez et al. 2016). These students conducted their doctoral research in cross-disciplinary teams of three to four members over the course of four or five years. Each team was recruited to work on socio-environmental issues in a specific location. The goal of each team was to identify an overarching research problem that it could investigate collectively, with each student writing a dissertation on an aspect of that problem and the team jointly authoring a paper that built on their collaboration. The purpose of these workshops was to help members of each team enhance their mutual understanding and coordinate their research perspectives as they developed their collaborative projects.

12.2.2 PARTICIPANTS

The participants in these parallel workshops were students in a range of environmental disciplines, including ecology, entomology, geology, hydrology, and sociology. Two teams had four members, and the third had three members. There were six men and five women. Some of the students had a moderate amount of experience conducting cross-disciplinary research, but most were new to it. Regardless of their prior experience, all of the students had chosen to pursue their doctoral degree in a cross-disciplinary program, so they were predisposed to appreciate the kind of cross-disciplinary exchanges that are the hallmark of Toolbox workshops.

12.2.3 INSTRUMENT DESIGN

We used the Scientific Research Toolbox instrument for these workshops (see Appendix A). The earliest version of this instrument (Eigenbrode et al. 2007) was produced with IGERT students in mind, and the version we used in this workshop had been developed specifically for IGERT implementation. Both the language of the prompts and the structure of the instrument were crafted to enable new scientific researchers to articulate the core beliefs and values that guided their research.

12.2.4 PROCESSES

Prior to the workshops, participants were asked to read Eigenbrode et al. (2007), Heemskerk et al. (2003), and an early version of Bennett and Gadlin (2012). In the workshops, the TDI team gave a presentation on the philosophical foundations of science and on the many cultures in science in order to prepare the participants for the experience. After this preamble, the three dialogue sessions were conducted in parallel, each lasting roughly 75 minutes. Participants then discussed conceptual models of the socio-environmental systems their teams would investigate. Finally, all teams met together to debrief on the experience.

12.2.5 OUTCOMES

Educational applications of Toolbox workshops like this one are one of the central activities for TDI. This is the kind of experience that motivated the development of the Toolbox dialogue method in the first place, and this particular set of workshops illustrates the positive influence of Toolbox workshops on cross-disciplinary graduate students early in their research training. The fact that the Toolbox approach itself emerged from an IGERT project similar to this one also resonated with the students and motivated buy-in and engagement. The dialogue also structured a follow-up activity that involved jointly developing and evaluating an initial concept map of the human and ecological components of their study area and study system. This concept map became a focal object for the teams over the subsequent year, receiving additional attention and undergoing modification as the teams developed their research projects.

As with many Toolbox workshops, we gathered three types of data. We collected rating-response data from the instruments before and after the dialogue, audio recordings of the dialogue (which were later transcribed), and post-workshop questionnaire responses. For each dialogue session, we assessed introspective effects manifested at the level of the individual, interactive effects manifested at the level of the teams, and evaluative effects reflected in the attitudes of the participants toward the Toolbox experience. We followed these teams through successful completion of their projects, and our impression is that the Toolbox experience was helpful in facilitating the teams' collaborative efforts.

From the perspective of *introspective* effects, participants reported that they learned about their teammates' perspectives, and several reported changes in how they think about aspects of science. All felt the workshops increased their desire to conduct cross-disciplinary research. Our assessment from the perspective of *interactive* effects is that the dialogue in each group was reflective, thoughtful, and engaged. It was clear that the participants were interested in learning about what each other had to say about the prompts. Participants generally agreed that the workshop showed the importance of mutual understanding of philosophical perspectives within the team. All of the dialogue sessions addressed questions of objectivity, values in science, and the centrality of hypotheses to the conduct of science. In addition to mutual teaching and learning, the teams also built connections during the sessions and displayed interest in and mutual respect for the comments made by their teammates.

From the perspective of *evaluative* effects, all three groups found the Toolbox dialogue useful. Participants noted the practical value of the dialogue as a structured way to initiate in-depth conversations. Participants also noted that the Toolbox could guide exploration of the importance of theory, methodologies, and hypotheses, thereby broadening the teams' understanding of the scientific enterprise.

12.3   REACCH GRADUATE STUDENT/POST-DOC RETREAT

In 2012, two Toolbox workshops were conducted as part of the Graduate Student/ Post-Doc Retreat for Regional Approaches to Climate Change (REACCH), a large, transdisciplinary, multi-institutional agriculture project funded by the US Department of Agriculture-National Institute of Food and Agriculture. TDI was a partner with REACCH over its entire lifecycle, contributing to its efforts to catalyze integrative and collaborative capacity (Salazar et al. 2012; Piso et al. 2016). This case study illustrates how the Toolbox dialogue method can be used to help graduate students reflect upon their individual and team research. It also demonstrates some characteristic responses that first-time workshop participants have to the TDI experience.

12.3.1   OBJECTIVES

The goal of these workshops was to prepare students for their research after joining REACCH. The workshops were part of a cross-disciplinary program of graduate education that featured student retreats and professional development. The retreat encouraged project-wide synergy and communication on both interdisciplinary and transdisciplinary aspects of the project. Specific aims of the retreat's Toolbox workshops included encouraging critical reflection and analysis and identifying values and priorities of both the group and its constituent members.

12.3.2   PARTICIPANTS

The two workshops were conducted in parallel during the retreat. One had 15 participants and the other had nine. All participants were either graduate students or post-doctoral fellows. All were conducting research related to the mission of REACCH, which was to ensure sustainable cereal production in a three-state region. A wide variety of disciplines were represented in these sessions, including agroecology, atmospheric science, ecology, economics, education, environmental science, geo-informatics, human performance technology, hydrology, plant pathology, sociology, and soil science.

12.3.3   INSTRUMENT DESIGN

These workshops used the Climate Science/Management Toolbox instrument. This instrument was one of the first designed specifically for a partner, having been

TABLE 12.1
The Theme and Core Question for Each of the Seven Modules in the Climate Science/Management Toolbox Instrument

Module Theme	Core Question
Motivation	What primarily motivates scientists to conduct climate science research?
Application	What role should researchers play in solving real-world problems?
Values-management	What values matter most in natural resource decisions?
Values-science	Do values have a legitimate role in scientific research?
Trust	Does relevant work on climate science require partnerships between scientists and natural resource managers?
Uncertainty	How does uncertainty affect climate science research and its application?
Ways of knowing	Must different ways of knowing be combined to adequately address climate issues?

previously developed in collaboration with representatives of a large, federally funded center in the United States devoted to climate science. Since the mission of REACCH also involved assessing the impacts of climate change, specifically on cereal production, TDI and the project leadership decided that it would be an effective instrument to use in these workshops.

The instrument contained seven modules. Their themes and core questions are found in Table 12.1. Illustrative prompts included in this instrument were "Bridging science and policy should be the principal contribution of climate science research" from the *Motivation* module, "Understanding social perceptions of climate change is as important for adaptation as understanding the biophysical aspects of climate change" from the *Application* module, and "Scientists should have greater respect for the complexity of natural resource management decisions" from the *Trust* module.

12.3.4 Processes

The retreat took place over two days. Participants arrived on a Saturday afternoon and dined together that evening; workshops were then run on Sunday morning. The preamble was delivered in the evening, with the dialogue portion of the workshop commencing first thing in the morning on Sunday. The workshop was intended to facilitate team building among the graduate students and post-docs and to provide the students with an opportunity to be more reflective about the practice of science and working in cross-disciplinary scientific teams.

12.3.5 Outcomes

REACCH worked with TDI in a number of different ways over the course of its lifecycle, beginning with a set of standard Toolbox workshops conducted during the project's launch meeting. A key commitment of this project was to cross-disciplinary training, and the Graduate Student Retreat was a central part of its training curriculum. TDI's experience with cross-disciplinary training for graduate students, as described in the previous IGERT case study (see Section 12.2), positioned us to contribute to graduate student training in REACCH. This exercise was repeated with modifications in the following year.

The workshop was well received by the students. In their feedback, they noted that workshop dialogue revealed differences in viewpoint that might have remained hidden without the Toolbox structure to motivate their disclosure. Discovering the variety of disciplinary paradigms available to a cross-disciplinary researcher when approaching a research question was a revelation for the students. Having gained this perspective, the students were better able to begin building a shared vision for their collaborative work together on the project. The dialogue also played an important educational role for these students by illuminating differences in how they thought about climate change and agriculture. It helped reinforce their common commitment to working on the issue. It allowed them to reflect on the challenge of maintaining scientific credibility when working with diverse stakeholders and pursuing research within a large interdisciplinary project. A subset of graduate students developed a REACCH-specific Toolbox instrument and administered it in a live "clicker"-enabled session for all project participants at a subsequent all-hands meeting. The REACCH-specific instrument included prompts addressing motivation for transdisciplinary researchers, measures of success in transdisciplinary projects, and issues around data sharing by collaborators.

12.4   WEST MICHIGAN CLIMATE RESILIENCY SUMMIT WORKSHOPS

This set of 13 Toolbox workshops was conducted in 2014 as part of a day-long Climate Resiliency Summit held in Grand Rapids, Michigan. These workshops were the culmination of a long process of Toolbox development that involved various partners interested in expanding regional capacity around systematic responses to climate change. These workshops also included the first we conducted in parallel without active facilitation. Their success highlights several aspects of the Toolbox process that have yet to be emphasized in the case studies, including the potential for meaningful research collaboration in designing a Toolbox workshop, the viability of philosophically based dialogue in a community context, and the scalability of the Toolbox dialogue method.

12.4.1   OBJECTIVES

There were two sources of motivation for these workshops. First, there was a growing desire among many individuals in Grand Rapids and the surrounding region to be proactive in the face of the changing climate. This desire was grounded in a close call the city had with a 2013 flood and the belief that more extreme weather events would occur. This led to a region-wide effort to examine the climate resiliency and sustainability of their infrastructure. Second, relationships with the Sustainable Climate Risk Management (SCRiM) project at Penn State University and the West Michigan

Sustainable Business Forum (WMSBF) in Grand Rapids made it possible for TDI to adapt the Toolbox dialogue method to the community context.

The primary objective for these Toolbox workshops was to enhance mutual understanding among members of different business sectors in West Michigan and build capacity to collaborate on a region-wide response to climate change. TDI was interested in using these workshops to evaluate a new way of developing a Toolbox instrument that involved values-informed mental models (Mayer et al. 2017) and to test how well the Toolbox dialogue method scaled in a context where many of the workshops would not be facilitated.

12.4.2   PARTICIPANTS

The 13 workshops involved 86 participants from the West Michigan region. Only three were academics. The rest included businesspeople, government energy and environment employees, government public health workers, non-governmental organization members, a professional advocate, and a solution provider. The participants were organized into 13 groups, two of which had facilitators. The two facilitated workshops included people who were identified by WMSBF as being key players in West Michigan on the issue of climate change.

12.4.3   INSTRUMENT DESIGN

The Toolbox instrument was developed in part using values-informed mental models from SCRiM to identify dialogue themes. The mental-models methodology involves using structured interviews to discover the conceptual models used by people to understand various topics, using an "expert" model as a baseline (Morgan et al. 2002). In this instance, the methodology was modified to highlight how an individual's values can affect understanding. The process focused on the mental models of 10 individuals selected by WMSBF to represent the diversity of regional stakeholders. The goal was to identify potentially controversial issues that were especially salient in this region. Once these issues were identified, the hope was that using dialogue to create common ground about them could benefit the future region-scale effort.

Analysis of the mental-models interviews yielded a set of four modules that encompassed the climate-related issues of greatest interest to the business and government sectors in West Michigan. (See Table 12.2 for the module themes and core questions.) Examples of prompts included in this instrument were "Any acceptable storm water management plan for West Michigan must not worsen flooding outside West Michigan" in the *Emergency Preparedness* module, "Environmental losses are permissible when offset by economic gains" in the *Sustainability* module, and "Those who profit the most from unsustainable energy practices have a special obligation to pay for mitigation and adaptation efforts" in the *Energy* module.

12.4.4   PROCESSES

The Climate Resiliency Summit was the climax of a series of events designed to mobilize residents of West Michigan around systematic responses to a changing

TABLE 12.2
The Theme and Core Question for Each of the Four Modules That Constitute the Climate Resiliency Toolbox Instrument

Module Theme	Core Question
Emergency preparedness	How should we prepare for emergencies and disasters in West Michigan?
Sustainability	How should West Michigan define sustainability in an uncertain future?
Energy	How should we approach energy production and consumption in West Michigan?
Government	How should government help West Michigan achieve climate resiliency?

climate. These events were sponsored by a number of Grand Rapids organizations, including WMSBF. Informational pre-meetings were utilized to increase understanding of climate science and the state of the climate in West Michigan. The pre-meetings were organized by sector, such as manufacturing, business, and the built environment. Each of the key informants participated in the informational pre-meeting that concerned their sector.

The summit was a day-long event that featured keynote speakers and the Toolbox workshops. The workshops were initiated in a plenary session by a TDI presentation that detailed the context and background of our work. The dialogue portion of the workshops then commenced. Nineteen participants gathered in two facilitated sessions in nearby breakout rooms, and 67 participants engaged in largely self-facilitated dialogue at tables in the ballroom. TDI had two people available to answer questions in the ballroom, although there was no active facilitation. After the dialogues concluded, the groups reconvened in the ballroom for a short debrief discussion with the facilitators.

12.4.5 OUTCOMES

This was the first significant foray by TDI into a non-academic context. It was largely successful. Responses to the post-workshop questionnaire supported our sense that participants appreciated the opportunity to have these conversations. Our primary regional partner, WMSBF, also found the Toolbox workshops to function as intended in the context of the Climate Resiliency Summit. The work we did with SCRiM to prepare for the workshops was more involved than typical, but it was revealing and ensured that we were responsive to the values and core beliefs of the people in the region. The result was an instrument that resonated with the participants and fostered substantial and informative exchanges in the dialogues. The sessions also demonstrated that our approach, which involves a focus on the participants and light facilitation, works even if there are dialogue groups that don't have dedicated facilitators.

Participants in the two facilitated sessions discussed a variety of important themes. They engaged in close conversation and at times expressed disagreement about core values related to climate change. Among the topics that generated debate in these sessions were whether education could be an effective vehicle for social change, whether government should play an expanded or limited role in securing climate resiliency for the region, and whether long-term investments in sustainable infrastructure make sound business sense. In both sessions, differences of opinion about important matters emerged that might not have been disclosed without being highlighted for consideration, and conversation about these differences involved interlocutors who might not otherwise have spoken to one another.

12.5   WOODY WEEDS PROJECT IN EAST AFRICA

In 2015, two TDI representatives virtually joined the second meeting of Woody Weeds (http://woodyweeds.org/), a natural resource management project set in East Africa. This was the second Toolbox workshop for cross-disciplinary researchers involved in this transdisciplinary and multinational project. Highlights of this case study are virtual delivery of the workshop, international participation, and long-term partner engagement.

Our partnership with Woody Weeds is an important, long-term relationship for TDI, and the workshop described in this case study was the second of five that we have conducted at the time of this writing. The "Context" section in this case study describes how this particular Toolbox workshop relates to the project as a whole.

12.5.1   Objectives

Woody Weeds is a project whose goal is to reduce the negative impact of certain woody invasive alien species in Kenya, Tanzania, and Ethiopia. It is funded by the Swiss Programme for Research on Global Issues for Development. Since the project's inception, TDI's role has been to help facilitate communication across disciplines and to help manage communication over the project's lifecycle. At the launch meeting held in Nairobi, Kenya, in 2015, all project members participated in a typical Toolbox workshop with prompts modified from the primary Scientific Research Toolbox instrument (see Appendix A). The workshop described in this case study took place at the second project-wide meeting, held in Amani, Tanzania, later in 2015. Dialogue in this workshop was structured by an instrument modified from the one used in the launch meeting to address project-specific concerns using the Solera method (see Chapter 11).

At the time of this workshop, participants were formulating the project's approach to the integration of information from genetics, hydrology, invasion ecology, management, remote sensing, and stakeholder engagement. The project leadership sought to help project members refine their understanding of the cross-disciplinary dimensions and challenges they were going to confront. The TDI facilitators aimed to build upon the progress made in the launch meeting by encouraging dialogue about themes linked to specific project challenges as identified by the participants.

12.5.2   Participants

There were two dialogue groups in this workshop, one with eight participants and the other with 10. Four participants were women. Six participants were doctoral students, three were early-career professionals, three were mid-career professionals, and six were late-career professionals. They came from the earth, environmental, life, and social sciences, and their home institutions were in four countries in Africa and in Switzerland.

12.5.3 INSTRUMENT DESIGN

Work on the instrument used in this workshop began during the prior project launch meeting. That meeting included a Toolbox workshop, a subsequent report of that workshop, and an exercise in which participants were guided in developing their own prompts related to the project. Subsequently, TDI representatives worked with volunteers from Woody Weeds to refine the new prompts and produce the instrument that was used in the Amani workshop. The new instrument had 36 prompts, 14 of which were new or modified. There were six modules, including a new one, *Organization* (Table 12.3). Examples of new prompts were "Curiosity-driven research has more academic value than applied research" in *Motivation*, "Environmental data represent reality and socio-economic data represent perception" in *Reality*, and "Designing scientific research around the interests of funding institutions impedes good science" in *Organization*.

12.5.4 PROCESSES

The workshop was conducted with all participants on site in Amani and the facilitators each connected by video link from different locations. In each dialogue session, the project leaders on site helped with logistics, recorded speaking turns, served as local moderators at the outset of the dialogue, and participated with the others in the dialogue. The meeting began with a plenary overview presentation from the facilitators that reviewed the activities and outcomes of the launch meeting workshops and how the new instrument had been created. This was followed by the dialogue sessions. Despite the potential difficulties associated with the partially virtual format, the

---

TABLE 12.3

The Theme and Core Question for Each of the Six Modules That Constitute the Toolbox Instrument Used in the Woody Weeds Workshop

Module Theme	Core Question
Motivation	Does the principal value of research stem from its applicability for solving problems?
Methodology	Are certain disciplinary methods more important for this project than others?
Reality	Do the products of scientific research more closely reflect the nature of the world or the researchers' perspective?
Values	Do values negatively influence scientific research?
Ways of knowing	Must different ways of knowing be combined to adequately address ecological issues?
Organization	How does organizational structure influence scientific research projects and innovation?

dialogue sessions went smoothly. A report was produced by the facilitators for the leadership that included a summary of the dialogues with highlights of the discussions of each prompt, a quantitative presentation of the pre/post-workshop rating-response scores generated from each prompt, and some recommendations for project leadership on how to build on the outcomes and improve the project.

12.5.5 OUTCOMES

Each dialogue session was collegial and friendly. Differences in research worldviews were evident, but the participants communicated effectively about these differences. The TDI facilitators noted that there were three important themes in both dialogue sessions: a divide in mutual appreciation between biophysical scientists and social scientists in the project, different opinions about the importance and place within the project of Indigenous or local knowledge, and a lack of understanding of what is required to successfully translate scientific results for policymakers. These themes became focal points in subsequent Toolbox workshops for the Woody Weeds project. Many of the most-discussed prompts were those developed specifically for this workshop, which indicates that involving participants in the crafting of prompts was effective.

This workshop identified and improved project-wide understanding of themes relevant to the Woody Weeds project. Feedback from the participants indicated that these outcomes helped them better understand the project and the issues and challenges to communication. For TDI, this workshop demonstrated that virtual workshops can be effective in supporting frank and revealing dialogues. It also demonstrated the value of working over a longer term with a collaborative project and providing continuous communication support as the project matures.

12.5.6 CONTEXT

This partnership is one of the most thorough and long-term relationships between TDI and a large transdisciplinary project. Table 12.4 illustrates the components of that relationship.

The issues identified in the launch-meeting workshop and in the All-Project Meeting workshop held in Amani, Tanzania were explored through various workshop designs in subsequent project meetings, enabling discussion of persistent issues and development of action steps to address them. The emphasis was increasingly on participant responsibilities and best practices to continue effective communication within the project. For example, in the subsequent Year Two meeting in Adama, Ethiopia, four small groups each worked on one of the four themes that had emerged as significant in the prior project-wide workshops: the relationship between scientific knowledge and Indigenous knowledge, methodological differences between qualitative and quantitative reasoning, effective communication and transparency in the project, and student integration into the project. In Year Four of the project, TDI methods were employed in a discussion with project stakeholders. Farmers, pastoralists, village leaders, and agency personnel working in the Afar region of Ethiopia were included. In addition to the workshops, TDI has conducted a short, annual "communication survey" of project participants to assess their views on communication

TABLE 12.4
Partnership Activities between TDI and the Woody Weeds Project. (The case study workshop took place during the All-Project Meeting in Year 1)

Year	Year 1 2015–2016	Year 1 2015–2016	Year 2 2016–2017	Year 3 2017–2018	Year 4 2018–2019	Year 4 2018–2019	Year 6 2020–2021
Meeting	Launch Meeting	All-Project Meeting	First Annual Project Meeting	Second Annual Project Meeting	Third Annual Project Meeting	All-Project Meeting	Final Project Meeting
TDI contribution	Two Toolbox workshops	Virtual Toolbox workshop	Toolbox workshop	Toolbox workshop	Toolbox workshop	Two Toolbox workshops	TDI assessment
Location	KEFRI, Nairobi, Kenya	Amani, Tanzania	Adama, Ethiopia	Nakuru, Kenya	Amani, Tanzania	Bishoftu, Ethiopia	TBD
Nature of workshop	STEM-based dialogue; identify communication issues and themes	Dialogue with students; instrument modified with input from project participants	Dialogue and action steps, core prompts	Dialogue and action steps, core prompts	Dialogue focused on execution	STEM-based dialogue with second cohort; dialogue with Ethiopian stakeholders	TBD

processes within the Woody Weeds project. Reports on workshops and on the communication survey have been regularly provided to project leadership, who have embraced the premise and approach of TDI and insisted on full participation in our activities. This has helped promote a culture of frank discussion, communication, and openness concerning challenges and differences in views about fundamental issues that affect project activities.

12.6 COMMUNITY LAND USE TOOLBOX WORKSHOP: WINNEBAGO COUNTY, ILLINOIS

In January of 2016, TDI moderated a Toolbox workshop for landowners and land managers from Winnebago County, Illinois, that involved dialogue about county agricultural lands among residents who were related in different ways to those lands. This case study demonstrates how the Toolbox dialogue method can be adapted to a non-academic, community context.

12.6.1 OBJECTIVES

This workshop was conducted to enable county residents engaged in agricultural practices to participate in dialogue about issues related to how county lands were used for agricultural purposes. It was facilitated as part of a research project conducted by Greg Keilback, a graduate student in the College of Natural Resources at the University of Idaho. He was interested in evaluating how county residents, especially farmers, perceived their environment and practices. His research involved assessing how the county's residents were adapting to changing political, social, and economic paradigms.

Keilback's project was motivated in part by the county's "2030 plan." The goal of the research was to promote policies for the use of agricultural land that would provide public benefits such as biodiversity conservation, soil conservation, and increased public health. These policies could include incentive-based programs like the Conservation Reserve Program that takes agricultural land and sets it aside for other uses (United States Department of Agriculture Farm Service Agency 2019). A key part of creating and fostering public benefits in this instance is the collection of public perceptions about land use practices and their shared landscape. The Toolbox workshop contributed dialogue data to this research effort. The workshop was intended to enable members of different groups that had close relationships with the land—e.g., environmentalists, farmers, landowners—to engage in dialogue about the use of agricultural lands in an ever-changing social and political landscape.

12.6.2 PARTICIPANTS

Workshop participants consisted of residents of Winnebago County who had lived there for at least 10 years. This restriction was imposed because it was important that the participants were familiar with agricultural lands and land-use practices in the county. The workshop included participants who were operators of industrialized, family-owned, and organic farms, and non-farm residents from rural and urban parts of the county. Participants were recruited from locations where community residents gathered—e.g., libraries and gas stations—and also by referral from county contacts. An effort was made to include a diversity of participants across a range of ages, races and ethnicities, length of residence in the county, and farming experience.

12.6.3   INSTRUMENT DESIGN

The instrument used in the Community Land Use Toolbox workshop included prompts that pertained to land, farming, and the environment. These prompts emphasized normative dimensions of land use and farming practice (e.g., values, priorities). They were more concrete than many Toolbox prompts. The instrument comprised three modules, with seven prompts collected under the question, "How often would you say that these statements are true?" (always – never scale), eight prompts under the question, "How much would you say that these statements are like you?" (a lot like me – not at all like me scale), and 11 prompts under the question, "How much do you agree or disagree with these statements?" (agree – disagree scale). Examples of prompts from the first module, "How often would you say that these statements are true?" included "I research and implement practices that will improve the health of my land (or land I control)" and "New advances in chemical application and genetic modification help create greater yields and revenue."

12.6.4   PROCESSES

TDI was able to deliver this workshop because of Keilback's connection with this county, which was his home. He approached TDI with an interest in running a Toolbox workshop as part of his Masters-degree research on practices involving the use of agricultural lands in a traditionally agricultural area that was undergoing significant change. After communicating with TDI representatives, he worked closely with a TDI member at the University of Idaho to design the instrument. The prompts that constitute the instrument used in the Toolbox workshop derived from Keilback's research and his experience. The TDI member joined Keilback on his research visit and facilitated the workshop.

The workshop started with a preamble delivered by the TDI facilitator. It emphasized the need to focus on finding common ground and avoiding negative stereotypes that would undermine communication. The dialogue was engaged, with lively and congenial exchanges that reinforced the importance of mutual awareness and a sense of community across the groups represented at the workshop.

12.6.5   OUTCOMES

The dialogue was an open exchange of ideas between representatives of groups that relate very differently to the land. The fact that this diverse group of county residents could converse for more than three hours demonstrates how much interest there was among county residents in participatory land-use planning. The participants were eager to learn about each other's perspectives on farming and land-use practices. Many of them contacted Keilback after the workshop to express the desire for more opportunities to learn from one another.

In addition to serving as an educational opportunity for the residents of the county, the workshop also provided Keilback with a number of research results. While it was clear that there were issues dividing the various groups, the data suggested that there were more similarities among them than differences. For example, the participants recognized the systematic interconnection of the land, and that this meant what one land user did had consequences for others in the community. The participants also expressed an appreciation for the intrinsic value of the land that went beyond its utility. A commitment to collective stewardship could be undone by negative stereotypes and mistrust, which underscored for the participants the importance of continually learning about the land and about each other. Finally, the group expressed suspicion about large, land-focused incentive programs run by corporations or government agencies, suggesting instead that conservation programs should be run by more local entities that had a connection to the land.

12.7   MICHIGAN STATE UNIVERSITY OFFICE OF SUSTAINABILITY

TDI conducted a workshop for the Michigan State University (MSU) Office of Sustainability in May 2016. The office was interested in gathering information from the university community about a sustainability planning process due to commence the following year. This case study describes a workshop designed for a non-research organization operating in a large institutional context.

12.7.1   Objectives

The workshop was structured by an instrument that was developed jointly by the Office of Sustainability and TDI. In addition to fostering conversation among university personnel about sustainability on campus, these workshops were intended to help the Office of Sustainability identify values and priorities related to campus sustainability projects. Information about these values and priorities could help them plan future sustainability programs. Workshop goals for the participants included assisting them with identifying values and priorities related to sustainability on campus, encouraging them to critically reflect on sustainability efforts at MSU as well as on previous sustainability processes, and helping them identify relevant stakeholders for sustainability efforts on campus. TDI used this collaboration to continue developing its ability to build collaborative capacity within non-research groups.

12.7.2   Participants

Participants were self-selected based on their interest in sustainability efforts at the university. They included faculty, graduate students, and staff from across MSU. Two parallel dialogue sessions took place. In each, 10 people participated in a dialogue that lasted approximately 90 minutes.

TABLE 12.5

The Theme and Core Question for Each of the Six Modules That Constitute the Toolbox Instrument Used in the MSU Sustainability Workshop

Module Theme	Core Question
Communication	How should MSU communicate with campus and the local community about sustainability?
Individual action	How should MSU incorporate individual stakeholders into campus sustainability planning?
Research	What role should research at MSU play in campus sustainability efforts?
Community interaction	What role should interaction with local communities play in MSU's campus sustainability efforts?
Campus sustainability	What should MSU's future campus sustainability goals look like?
Education	How should MSU include education efforts in campus sustainability planning?

12.7.3 INSTRUMENT DESIGN

The instrument was designed collaboratively with leadership from the MSU Office of Sustainability. After studying campus sustainability documents and meeting with campus officials, TDI representatives drafted initial instrument language and then shared that with the director of the office. After an iterative process of prompt refinement and selection, six modules were selected to constitute the instrument (see Table 12.5).

The prompts in this instrument were designed to focus on the sustainability issues confronting MSU. Since the context of the workshop was a university, research and educational aspects of sustainability were priorities. The Office of Sustainability was also keen to learn about the willingness of individuals and of the community to engage in action conducive to sustainability. Finally, it was determined that communication would be a key part of the university response to sustainability, no matter how it played out, and so communication figured importantly in the final instrument. Examples of prompts from the *Communication* module included "MSU's sustainability plans for the future are not as clear as they should be" and "MSU effectively communicates its energy strategy to the campus community," while the *Individual Action* module contained prompts such as "Community members should do more to influence campus sustainability efforts" and "All individuals on campus should be held accountable for their energy costs."

12.7.4 PROCESSES

The workshop began with a 15-minute presentation by the facilitators in which they described the design and purpose of the exercise. The participants were then split into two groups for the dialogue sessions. Upon completing the dialogue and the instrument, participants reassembled and were asked to discuss the themes that were raised in their dialogue session. Common themes included the nature of incentives for sustainability practice, what is needed to convince individuals to change their behavior, and the lack of clarity about the concept of *sustainability*. After this discussion, participants were divided into groups of four or five participants. Members of the two dialogue sessions were mixed. Each group was asked to devise five messages for the Office of Sustainability based on the Toolbox dialogue. This activity confirmed that communication, definitions, and campus culture were important for many of the participants. A follow-up survey was sent to workshop participants 10 days after the workshop.

12.7.5 OUTCOMES

This workshop demonstrates the power of Toolbox dialogue to focus members of a large community on topics important for that community. The participants in the workshop exhibited a deep passion and interest in discussing the university's sustainability efforts. It represented an opportunity for a diverse group of people with a common interest in sustainability to gather together and think in a focused way about it, working out their specific views and collectively planning a way forward. The Toolbox intervention also created a space for dialogue to inform the MSU Office of Sustainability about stakeholder perspectives, values, ideas, and needs. While the workshop included a diverse range of university stakeholders, it could have been even more diverse; for example, there were no undergraduate students in attendance, even though undergraduates had been active in championing sustainability on this campus. Although this was the only workshop conducted for the Office of Sustainability, commitment to running a series of workshops like this could have had the effect of mobilizing campus citizens in support of a more participatory sustainability planning process.

The workshop dialogue generated insights that were useful for the Office of Sustainability. One insight concerned the need to think about sustainability broadly rather than limiting it to energy consumption. There was also confusion in both dialogue groups about the notion of *sustainability*. Although the size and scope of units on this campus doing sustainability-focused work varied considerably, we argued in our workshop report that it could foster cohesion and buy-in to create an inclusive definition of *sustainability* for the university in which community members could find representation of their efforts. Additionally, creating compelling campus initiatives tied to a university definition of *sustainability* could increase community engagement and also influence individual actions. Participants in both dialogue sessions expressed a desire for deeper university engagement with the local community, which could take various forms, such as work with local scouts or service-based classes on campus. Incentives and individual actions were topics of discussion in both workshops. We argued in the report that future campus sustainability planning should focus in part on influencing individual actions, perhaps through the use of incentives.

12.8 DEVELOPMENT FOR A NON-GOVERNMENTAL ORGANIZATION

TDI conducted a workshop for Catholic Relief Services (CRS), a large, international development NGO in October 2016. This workshop was the first one for TDI that had the explicit goal of building collaborative capacity in organizations that focused on global development and humanitarian aid, and it was also the first we conducted with an NGO. It exemplifies how the Toolbox dialogue method can be adapted for new thematic and organizational contexts.

12.8.1 Objectives

CRS works to address and remediate hunger, poverty, and disease. In doing so, it draws upon people from different nations, cultures, and professional sectors. These collaborators can struggle to find common ground. To this extent, it is an organization in which Toolbox-style dialogue can help collaborators recognize implicit differences that influence attitudes about development projects and the business of development more generally. This Toolbox workshop was offered to CRS, as a first attempt at extending the Toolbox dialogue method to the global development sector. After working for six months to adapt the method to the development context and identify potential partners, we approached CRS and offered to conduct a Toolbox workshop for members of their operation as a demonstration of its ability to reveal and coordinate implicit differences in values and beliefs. They welcomed the opportunity to have a structured conversation about core beliefs and values concerning development. They suggested that we conduct the workshop with a team that works on capacity-building within the organization. We believed that the workshop would have value for CRS and that it would also help TDI better understand how Toolbox dialogue could help to address the challenges and difficulties that arise within global development.

12.8.2 Participants

The eight participants worked in the internal unit that focused on providing knowledge and training to support capacity-building within CRS. Some worked in the main office in the United States, while others came from some of the international offices. The participants knew one another and this familiarity helped the group generate engaged dialogue. The professionals working in this unit were highly skilled in areas such as planning, monitoring, and evaluation processes since they contributed to training efforts for their colleagues, members of other NGOs, partner organizations, and others. These efforts included skills training, as well as mentoring and other forms of support for individuals and organizations.

12.8.3 Instrument Design

The dialogue was structured by a four-module Toolbox instrument. The modules and core questions are shown in Table 12.6. This instrument was designed by TDI members, including one with experience in the global development sector. The prompts were designed to probe how the participants conceived of development, the sustainability of development projects, and normative aspects of development practice, including justice and understanding. Example prompts included "Collaborators working on a development project should have a shared conception of development" from the *Development* module, and "Development projects concerned with justice

TABLE 12.6
The Theme and Core Question for Each of the Four Modules That Constitute
the Toolbox Instrument Used in the Global Development Workshop

Module Theme	Core Question
Development	How important is understanding the concept of development for development practice?
Justice	How do differing understandings of justice influence how injustices are addressed?
Knowledge/cultural understanding	How important is it to development to consider what people take knowledge to be?
Sustainability	How should sustainability be understood in a development context?

must address aspects of injustice to individuals (e.g., marginalization, discrimination, powerlessness)" from the *Justice* module.

12.8.4 PROCESSES

No pre-workshop engagement took place with this group apart from email exchanges with the director of the capacity-building unit. The workshop itself began with a preamble by the facilitator, followed by a substantial and lengthy dialogue. The instrument was completed both immediately before and immediately after the workshop. The workshop concluded with a substantial debrief about the process. The TDI representatives then talked with the participants during lunch about the practice of development work.

12.8.5 OUTCOMES

The dialogue revealed that many of the issues explored in the instrument were central to the worldviews of the participants. It also indicated that dialogue of this sort could enhance the functionality of development teams. Issues that received attention in the dialogue included whether sustainability requires equal moral standing of humans and non-humans, whether the practice of development is too often limited by distributive conceptions of justice, and whether the biases of participants in development projects should be managed. Participants also wished to address the nature and impact of power dynamics. For example, questions were raised about whether development projects could be successful if the distribution of power wasn't acknowledged, and whether power imbalances underlie the structural injustice that is common among development projects.

The participants were engaged and thoughtful throughout the dialogue. We believe this reflects the fact that the issues raised in the dialogue are central to the practice of international development. The participants exhibited an impressive amount of analysis and reflection, which likely is a consequence of the institutionalized practice of reflection that is part of CRS culture. Participants were relaxed during the dialogue and allowed each other to express complex thoughts. There was laughter and some talkover, which indicated interest and excitement. It was clear to the TDI representatives that this type of dialogue could increase mutual understanding and communicative capacity for development teams that are new or involve members who are not familiar with one another. It would be interesting to use the workshop to compare cultures across development organizations, especially insofar as they conceptualize core aspects of development practice. As we noted in Chapter 11, we have continued our work in this sector, and are in the process of spreading the word about this form of capacity-building across our expanding network of development contacts.

12.9   SMALL RESEARCH TEAM WORKSHOPS

In 2017, TDI conducted five workshops at Purdue University, a large, research-intensive, land grant university. Four of the workshops were conducted for small, collaborative projects funded internally by two centers that had received a large foundation grant intended to aid the integration of humanities and social science into cross-disciplinary research on grand challenges, such as climate change and world hunger. In addition, a fifth workshop was conducted for the team that directed the foundation-funded project. These workshops included dialogue sessions designed for the teams, along with co-creation activities aimed at enabling the teams to convert what they learned in the dialogue into a deliverable they could take with them (e.g., a project glossary). They are good examples of the utility of Toolbox interventions at the beginning of the lifecycle of cross-disciplinary teams.

12.9.1   OBJECTIVES

Two academic centers at Purdue University received a foundation grant to support cross-disciplinary research teams working on grand challenges that included humanities and social science faculty. In addition, the grant aimed to improve scholarly communications for policy and social impact. The centers served as the sub-grantor for the award and held a competition within the university for the sub-awards. Four teams were selected. Each team had to include a representative from the humanities and a librarian in their proposal. As a sub-grantor, the centers also provided support for the teams in the form of capacity-building workshops focused on cross-disciplinary collaboration and on translating research for the public and policymakers. These workshops were commissioned by and involved the Principal Investigators (PIs) for the foundation grant.

All of the teams were in the early stages of their projects, having just received an award a few months before the Toolbox workshop. None of them had ever worked together as a team, although some individuals had worked together on other projects in the past.

12.9.2   PARTICIPANTS

The workshop teams varied in size from three to 10 people and included faculty and library staff at various career stages. Two teams also had graduate students attend. Some teams had members who did not participate in the workshop. Participants represented a range of disciplines including arts and humanities, communication science, computer science, engineering, library and informational sciences, life sciences, and social and behavioral sciences.

12.9.3 INSTRUMENT DESIGN

The instrument for these workshops was a customized instrument with three modules. Two of the modules, *Interdisciplinarity* (core question: "What are the specific challenges of doing interdisciplinary research?") and *Communication and Actionable Research* (core question: "How must we communicate our research to policymakers and other non-researchers in order to have the greatest impact?"), were included in the instrument for all teams, with a few prompts within these modules tailored to each team's area of research. The third module led with the core question, "What are the conceptual dimensions of our project?" and was tailored to each team's research theme, e.g., big data ethics, avoiding climate tipping points, and agricultural sustainability.

The instrument's prompts were a mix of those from previous instruments and some new prompts. The PIs for the foundation grant were invited to review the prompts. After this review, TDI staff finalized the instruments for each team. Examples of prompts found in the *Communication and Actionable Research* module included "Policy briefs are a more important communication mechanism than academic articles for our project" and "Scientists have as much to learn from policymakers as policymakers have to learn from scientists."

12.9.4 PROCESSES

Prior to the workshop, the PIs of the two centers completed a questionnaire to help TDI staff understand the goals for the proposed workshop, possible challenges facing the teams they were funding, current obstacles to communication and collaboration, general details about the initiative, and logistics for the workshops. A brief questionnaire was also distributed to the members of each team. These team questionnaires identified whether the respondent had previously worked with members of the team, had previous experience working on interdisciplinary teams (and, if so, what lessons they learned), had concerns about interdisciplinary projects, had experienced challenges to communication or collaboration, and had individual goals for the workshop. This information was helpful in planning the workshops, particularly in developing the instrument and co-creation activities.

12.9.5 OUTCOMES

Each workshop team participated in a co-creation activity. These activities varied slightly among teams. For the four research teams, the co-creation activity was designed to identify key concepts for the projects and start a team glossary with shared meanings for the concepts identified. This activity was facilitated in one of two ways. In the first, participants took a break from speaking in order to write one key concept and its definition on a piece of paper. Each participant passed it to their right. The recipient was invited to add to or question the concept and definition as well as to add another concept. After one or two rounds, the results were then shared with the larger group. The second sort of facilitation involved participants first brainstorming a list of concepts on their own, then sharing the list with a partner, and finally sharing the results with the larger group. After both kinds of co-creation activity, there was a facilitated discussion in which the concepts identified were presented for the whole group to evaluate, refine, and possibly supplement. The co-creation activity for the workshop involving the team that directed the foundation-funded project had a similar format. The focus was on individual, internal, and external goals. This was done to assist the team in identifying and prioritizing the goals of multiple actors on the team. The activity generated a long list of goals that were later consolidated and put into an online survey to allow team members to rank them in order of priority. All of the information collected from the co-creation activities for each group was emailed to the participants within a week after participating in the workshop so that the team members could continue the conversation while it was still fresh in their minds.

Responses to the feedback questionnaire indicated that participants enjoyed the Toolbox workshop and felt it had a positive influence on their team as a whole. Participants agreed that the experience was helpful in increasing understanding of their project's key concepts and/or goals and that the prompts helped stimulate conversation and identify different ways of thinking about their project. They reported a number of key insights that were related to team members' expectations and group dynamics, the importance of cross-disciplinarity and the role that disciplinary training plays in the use of terms or concepts, and the challenges the teams will face. A few participants expressed the need for more time to examine project-specific goals, to outline roles, and to make future plans.

This experience was atypical for TDI in that during the planning of the workshops, we were in contact only with the PIs for the two centers that served as the sub-grantor and not with the PIs of each research team. This resulted in communication challenges. In hindsight, we should have requested that the PIs of each research team be included in planning efforts from the beginning, which could have possibly garnered more buy-in from teams and individual team members.

12.10   ACKNOWLEDGMENTS

We thank all of the partner groups who participated in the workshops described in this chapter, and especially the leaders of these groups who worked closely with us to design and deliver the workshops. We thank other TDI personnel who were directly involved in developing and delivering these workshops: Nilsa Bosque-Pérez, Chad Gonnerman, Troy E. Hall, Brian Robinson, Stephanie E. Vasko, and J.D. Wulfhorst. For their contributions to these workshops, we also thank Bryan Cwik, René Eschen, Leontina Hormel, Jodi Johnson-Maynard, Greg Keilback, Klaus Keller, Penny Morgan, Zachary Piso, Urs Schaffner, Daniel Schoonmaker, Nancy Tuana, Sean Valles, and Kyle Whyte. Finally, we thank Steven Orzack for his editorial assistance. Eigenbrode's work with REACCH was supported by USDA NIFA Award #2011-68002-30191. O'Rourke's work on this chapter was supported by the USDA National Institute of Food and Agriculture, Hatch Project 1016959.

12.11   LITERATURE CITED

Bennett, L. M., and H. Gadlin. 2012: Collaboration and team science: from theory to practice. Journal of Investigative Medicine 60:768–775.

Berling, E., C. McLeskey, M. O'Rourke, and R. T. Pennock. 2019: A new method for a virtue-based responsible conduct of research curriculum: Pilot test results. Science and Engineering Ethics 25:899–910.

Bosque-Pérez, N. A., P. Z. Klos, J. E. Force, L. P. Waits, K. Cleary, P. Rhoades, S. M. Galbraith, A. L. B. Brymer, M. O'Rourke, S. D. Eigenbrode, B. Finegan, J. D. Wulfhorst, N. Sibelet, and J. D. Holbrook. 2016: A pedagogical model for team-based, problem-focused interdisciplinary doctoral education. BioScience 66:477–488.

Eigenbrode, S. D., M. O'Rourke, J. D. Wulfhorst, D. M. Althoff, C. S. Goldberg, K. Merrill, W. Morse, M. Nielsen-Pincus, J. Stephens, L. Winowiecki, and N. A. Bosque-Pérez. 2007: Employing philosophical dialogue in collaborative science. BioScience 57:55–64.

Hall, T. E., Z. Piso, J. Engebretson, and M. O'Rourke. 2018: Evaluating a dialogue-based approach to teaching about values and policy in graduate transdisciplinary environmental science programs. PLOS ONE 13:e0202948.

Heemskerk, M., K. Wilson, and M. Pavao-Zuckerman. 2003: Conceptual models as tools for communication across disciplines. Conservation Ecology 7:8.

Hessels, A. J., B. Robinson, M. O'Rourke, M. D. Begg, and E. L. Larson. 2015: Building interdisciplinary research models through interactive education. Clinical and Translational Science 8:793–799.

Kjellberg, P., M. O'Rourke, and D. O'Connor-Gomez. 2018: Interdisciplinarity and the undisciplined student: Lessons from the Whittier Scholars Program. Issues in Interdisciplinary Studies 36:34–65.

Knowlton, J., K. Halvorsen, R. Handler, and M. O'Rourke. 2014: Teaching interdisciplinary sustainability science teamwork skills to graduate students using in-person and web-based interactions. Sustainability 6:9428–9440.

Mayer, L. A., K. Loa, B. Cwik, N. Tuana, K. Keller, C. Gonnerman, A. M. Parker, and R. J. Lempert. 2017: Understanding scientists' computational modeling decisions about climate risk management strategies using values-informed mental models. Global Environmental Change 42:107–116.

Morgan, M. G., B. Fischhoff, A. Bostrom, and C. J. Atman. 2002: Risk Communication: A Mental Models Approach. Cambridge University Press, Cambridge.

Piso, Z., M. O'Rourke, and K. C. Weathers. 2016: Out of the fog: Catalyzing integrative capacity in interdisciplinary research. Studies in History and Philosophy of Science Part A 56:84–94.

Salazar, M. R., T. K. Lant, S. M. Fiore, and E. Salas. 2012: Facilitating innovation in diverse science teams through integrative capacity. Small Group Research 43:527–558.

Schnapp, L. M., L. Rotschy, T. E. Hall, S. J. Crowley, and M. O'Rourke. 2012: How to talk to strangers: facilitating knowledge sharing within translational health teams with the Toolbox dialogue method. Translational Behavioral Medicine 2:469–479.

United States Department of Agriculture Farm Service Agency. 2019: *Conservation Reserve Program: Fact Sheet*. Downloaded from www.fsa.usda.gov/Assets/USDA-FSA-Public/usdafiles/FactSheets/2019/conservation-reserve_program-fact_sheet.pdf.

Appendix A
Scientific Research
Toolbox Instrument

A.1 INTRODUCTION

In this appendix, we present the Scientific Research Toolbox instrument, which TDI has used in over 100 Toolbox workshops. It is also known as the "STEM" Instrument, for "Science-Technology-Engineering-Mathematics." Toolbox dialogues based on this instrument are the source of the quantitative and qualitative evidence discussed in Chapters 9 and 10. All of the Toolbox dialogue threads presented in Appendix B are based on this instrument. It was originally published by Looney et al. (2014). We thank SAGE Publishing for permission to present it here. This instrument was developed in 2008 and 2009 by a team led by Shannon Donovan, Sanford Eigenbrode, and Michael O'Rourke, with support from the National Science Foundation (NSF SES-0823058).

A.2 LITERATURE CITED

Looney, C., S. M. Donovan, M. O'Rourke, S. J. Crowley, S. D. Eigenbrode, L. Rotschy, N. A. Bosque-Pérez, and J. D. Wulfhorst. 2014: Seeing through the eyes of collaborators: Using Toolbox workshops to enhance cross-disciplinary communication. Pp. 220–243 *in* M. O'Rourke, S. J. Crowley, S. D. Eigenbrode, and J. D. Wulfhorst, eds. Enhancing Communication and Collaboration in Interdisciplinary Research. SAGE, Thousand Oaks, CA.

Motivation

Core Question: Does the principal value of research stem from its applicability for solving problems?

1. The principal value of research stems from the potential application of the knowledge gained.

   Disagree   *Agree*
        1   2   3   4   5   I don't know   N/A

2. Cross-disciplinary research is better suited to addressing applied questions than basic questions.

   Disagree   *Agree*
        1   2   3   4   5   I don't know   N/A

3. My disciplinary research primarily addresses basic questions.

   Disagree   *Agree*
        1   2   3   4   5   I don't know   N/A

4. The importance of our project stems from its applied aspects.

   Disagree   *Agree*
        1   2   3   4   5   I don't know   N/A

5. The members of this team have similar views concerning the motivation core question.

   Disagree   *Agree*
        1   2   3   4   5   I don't know   N/A

Methodology

Core Question: What methods do you employ in your disciplinary research (e.g. experimental, case study, observational, modeling)?

1. Scientific research (applied or basic) must be hypothesis driven.

   Disagree   _Agree_
   1   2   3   4   5   I don't know   N/A

2. In my disciplinary research, I employ primarily quantitative methods.

   Disagree   _Agree_
   1   2   3   4   5   I don't know   N/A

3. In my disciplinary research, I employ primarily qualitative methods.

   Disagree   _Agree_
   1   2   3   4   5   I don't know   N/A

4. In my disciplinary research, I employ primarily experimental methods.

   Disagree   _Agree_
   1   2   3   4   5   I don't know   N/A

5. In my disciplinary research, I employ primarily observational methods.

   Disagree   _Agree_
   1   2   3   4   5   I don't know   N/A

6. The members of this team have similar views concerning the methodology core question.

   Disagree   _Agree_
   1   2   3   4   5   I don't know   N/A

Confirmation

Core Question: What types of evidentiary support are required for knowledge?

1. There are strict requirements for the validity of measurements.

    Disagree   _Agree_
          1   2   3   4   5   I don't know   N/A

2. There are strict requirements for determining when empirical data confirm a tested hypothesis.

    Disagree   _Agree_
          1   2   3   4   5   I don't know   N/A

3. Validation of evidence requires replication.

    Disagree   _Agree_
          1   2   3   4   5   I don't know   N/A

4. Unreplicated results can be validated if confirmed by a combination of several different methods.

    Disagree   _Agree_
          1   2   3   4   5   I don't know   N/A

5. Research interpretations must address uncertainty.

    Disagree   _Agree_
          1   2   3   4   5   I don't know   N/A

6. The members of this team have similar views concerning the confirmation core question.

    Disagree   _Agree_
          1   2   3   4   5   I don't know   N/A

Reality

Core Question: Do the products of scientific research more closely reflect the nature of the world or the researcher's perspective?

1. Scientific research aims to identify facts about a world independent of the investigators.

   Disagree   *Agree*

     1   2   3   4   5   I don't know   N/A

2. Scientific claims need not represent objective reality to be useful.

   Disagree   *Agree*

     1   2   3   4   5   I don't know   N/A

3. Models invariably produce a distorted view of objective reality.

   Disagree   *Agree*

     1   2   3   4   5   I don't know   N/A

4. The subject of my research is a human construction.

   Disagree   *Agree*

     1   2   3   4   5   I don't know   N/A

5. The members of this team have similar views concerning the reality core question.

   Disagree   *Agree*

     1   2   3   4   5   I don't know   N/A

Values

Core Question: Do values negatively influence scientific research?

1. Objectivity implies an absence of values by the researcher.

    Disagree   *Agree*

    1   2   3   4   5   I don't know   N/A

2. Incorporating one's personal perspective in framing a research question is never valid.

    Disagree   *Agree*

    1   2   3   4   5   I don't know   N/A

3. Value-neutral scientific research is possible.

    Disagree   *Agree*

    1   2   3   4   5   I don't know   N/A

4. Determining what constitutes acceptable validation of research data is a value issue.

    Disagree   *Agree*

    1   2   3   4   5   I don't know   N/A

5. Allowing values to influence scientific research is advocacy.

    Disagree   *Agree*

    1   2   3   4   5   I don't know   N/A

6. The members of this team have similar views concerning the values core question.

    Disagree   *Agree*

    1   2   3   4   5   I don't know   N/A

Reductionism

Core Question: Can the world under investigation be reduced to independent elements for study?

1. Differences in spatiotemporal scales impede useful synthesis in cross-disciplinary research.

    Disagree   _Agree_
        1   2   3   4   5   I don't know   N/A

2. The world under investigation is fully explicable as the assembly of its constituent parts.

    Disagree   _Agree_
        1   2   3   4   5   I don't know   N/A

3. The world under investigation must be explained in terms of the emergent properties arising from the interactions of its individual components.

    Disagree   _Agree_
        1   2   3   4   5   I don't know   N/A

4. My research typically isolates the behavior of individual components of a system.

    Disagree   _Agree_
        1   2   3   4   5   I don't know   N/A

5. Scientific research must include explicit consideration of the environment in which it is conducted.

    Disagree   _Agree_
        1   2   3   4   5   I don't know   N/A

6. The members of this team have similar views concerning the reductionism core question.

    Disagree   _Agree_
        1   2   3   4   5   I don't know   N/A

	Demographic Profile (ID# _____)
1	Gender _____
2	Career phase: Early _____ (1–7 yrs) Mid _____ (8–20 yrs) Late _____ (20+ yrs)
3	Number of years you have been participating in inter- or cross-disciplinary activities: _____ yrs
4	What discipline(s) or profession(s) would you describe as your primary identity? 1. _____ 3. _____ 2. _____ 4. _____
5	Ethnicity: _____

Is there anything else that you would like to share with us?

Appendix B
Examples of Toolbox Dialogue

B.1 INTRODUCTION

In this appendix we present examples of dialogue from six Toolbox workshops. These examples represent the kinds of dialogue that arise when scientific researchers reflect on their research assumptions and share them in ways that are structured by philosophical prompts. We have selected six examples that provide the broader context of the transcript excerpts analyzed in Chapters 7 and 10.

These examples were taken from Toolbox dialogues that were transcribed verbatim and then annotated. These annotations include filler words (e.g., "umm," "uhh") and pauses (either explicitly in square brackets or with ellipses). Curly brackets, "{ }," denote an attempt by the transcriber to identify a spoken word or phrase. Square brackets, "[]," contain either a transcriber comment (e.g., "unclear," "can't hear"), an expansion of an acronym introduced for clarity, or a property of the dialogue that cannot be transcribed in words (e.g., continuation of speaking turns, overlap, crosstalk, laughter, mixed comments, talkover). Overlap, crosstalk, and talkover indicate a high level of participant engagement in the dialogue.

TDI members process each annotated transcript as follows. The audio and the transcription are compared and any transcription errors are corrected. Group and individual names are removed, with participants identified as "[P1]," "[P2]," etc. and names of non-participants mentioned in dialogue changed to "[name]." Each speaking turn is assigned to a participant.

A processed transcript is analyzed by having multiple readers assign the speaking turns to thematic clusters or "threads." Each of the examples below is a thread as determined by two TDI readers. A typical transcript will have many threads, which are usually initiated by explicit consideration of a new prompt. In some cases, the participant group will discuss one prompt at a time, resulting in as many threads as prompts discussed. However, participants will sometimes discuss multiple prompts in a single thread, as in examples 1, 2, 4, and 5 below.

Each thread is numbered according to its "Dialogue session" (DS) as indicated in Table 10.1 in Chapter 10. After noting the specific excerpt from Chapter 7 or Chapter 10 associated with the thread, we describe the participant group and the prompt(s) discussed. For each participant, we list their self-identified primary discipline, on the assumption that this is a core part of their professional identity. The columns in the tables correspond to the speaking turn number in the dialogue overall, the participant number of the speaker ("F" indicates the facilitator), and the content of the speaking turn.

B.2   THREAD FROM DS1

The thread in Table B.1 is associated with the excerpt discussed in Chapter 7 and Excerpt 10.4 from Chapter 10. The participants were members of a governing body that planned activities and allocated research support resources for an academic center focused on sustainability. The primary discipline for each participant (P) was as follows: P1 (chemical engineering), P2 (materials chemistry), P3 (mechanical engineering), P4 (chemistry), P5 (ecology), and P6 (economics). No prompt was explicitly discussed, but the thread concerns issues that are related to the following prompts, numbered according to their appearance in the instrument in Appendix A:

METHODOLOGY 1: Scientific research (applied or basic) must be hypothesis driven.

METHODOLOGY 5: In my disciplinary research, I employ primarily observational methods.

METHODOLOGY 6: The members of this team have similar views concerning the methodology core question.

CONFIRMATION 2: There are strict requirements for determining when empirical data confirm a tested hypothesis.

CONFIRMATION 3: Validation of evidence requires replication.

CONFIRMATION 4: Unreplicated results can be validated if confirmed by a combination of several different methods.

TABLE B.1
A Toolbox Dialogue Thread Extracted from DS1 That Addresses Issues Related to Scientific Methodology and Confirmation

Speaking Turn No.	Participant	Speaking Turn
157	P5	So do you believe that if you can't do the experiment, it's not science? [unclear]
158	P2	Well, it depends what you mean by "experiment." [talkover] I think there are … experiments are … I would say what you do in looking for patterns has an element of experiment in it.
159	P1	[overlap] Absolutely.
160	P2	[continued] You go out, you observe; you look for patterns. It's an observational experiment. So experimentation and observation are two pieces of science.
161	P6	So it doesn't require replication, or it does require replication?
162	P2	Oh, I think for … I would say, in the science that I do, replication is important.

TABLE B.1

A Toolbox Dialogue Thread Extracted from DS1 That Addresses Issues Related to Scientific Methodology and Confirmation (Continued)

Speaking Turn No.	Participant	Speaking Turn
163	P6	Right, but in other science, [can't hear]?
164	P2	Oh. Well, that's … I don't know.
165	P1	[overlap] Well, there's other ways to replicate, right?
166	P2	I don't really know what we're …
167	P4	[overlap] Sometimes you just can't replicate.
168	P2	[continued] I'd have to be given a "for instance," probably, to be able to test that; I don't—
169	P1	[overlap] Let's look at evidence of climate change, going back and looking at ice cores or other information. I mean, the replication is not necessarily actually performing experiments using comparable, sort of, spatiotemporal sets of data and trying to see coincidence of …. There was a …. You know, I mean, there were good examples of this. I was watching this Nova show on super volcanoes—you can imagine why, right? [laughter]—but it pointed to three different events that were occurring. Three different measurements that were made of post data, and they all came together and it was clear that this thing was being … one was looking at the chemistry of the ash; the other one was looking at sort of the topology of the structure of the earth; I forget what the third one was. But it was completely independent work, but replication … it's not true replication [name], but I mean, it's sort of validation in the sense that all three were coincident on exactly the same time period when this giant volcano went off.
170	P6	So we want to understand and prevent the recurrence of the Rwandan genocide.
		[talkover]
171	P1	[overlap] Yeah.
172	P6	[continued] That's a unique event.
173	P1	[overlap] How do you do it?
174	P6	[continued] You certainly don't want to replicate it, right? So … can you do science on—
175	P2	[overlap] That's right, and there are some things where you wouldn't want to set up the experiment for moral reasons.
176	P1	[overlap] What's the cause and effect?
177	P6	[overlap] Right, so would you … can you do science on the—

(continued)

TABLE B.1
A Toolbox Dialogue Thread Extracted from DS1 That Addresses Issues Related to Scientific Methodology and Confirmation (Continued)

Speaking Turn No.	Participant	Speaking Turn
178	P1	[overlap] Well, isn't that sociological research, to go back and see what other events ... we've had other horrific things, you know, in Cambodia and—
179	P6	[overlap] right, yeah, just picking an example.
180	P1	[overlap] Yeah. So you look at all of them and say 'What was your—
181	P6	[overlap] ... The point is, it's not replicable.
182	P1	[overlap] But there's root causes to that, that ... you know—
183	P?	[overlap] comparative.
184	P2	[overlap] I see what you mean.
185	P?	[overlap] It's not replication [can't hear].
186	P2	[overlap] In fact, even the first time is not an experiment. It's an event or a series of events that has an outcome, and you're just observing it.
187	P?	[overlap] Right.
188	P2	[continued] So I don't think you can set that up as an ... you don't ... it's not set up originally as an experiment. There's no controls, as near as I know; you're just observing it.
189	P6	[overlap] Right.
190	P2	[continued] I don't think you could even ... even if you wanted—morals aside—to replicate it, that you could. So I—
191	P6	Yeah.
192	P6	[overlap] But can you do science on it, is the question.
193	P5	[overlap] [can't hear] {for} the cause.
194	P3	[overlap] Is it opinion or is it [can't hear]?
		[mixed voices, talkover]
195	P2	[overlap] I don't know whether they call it sciences, but—
196	P3	[overlap] Somebody writes a paper on it; they say [can't hear] ...
197	P2	[overlap] ... it's social ... I think that social science and history and whatever, I think all those pieces have a component. It's not like physical. It's not physical science; that's clear.
198	P6	Yeah.
199	P4	But is it science?
		[laughter]
200	P6	{You're doing} a great job dodging [can't hear].

TABLE B.1

A Toolbox Dialogue Thread Extracted from DS1 That Addresses Issues Related to Scientific Methodology and Confirmation (Continued)

Speaking Turn No.	Participant	Speaking Turn
		[laughter, mixed comments, talkover]
201	P2	So let me say that I would put it in the realm of scholarly activity, but I—
		[laughter]
202	P2	[continued] Scholarly activity is not all research.
203	P?	[overlap] Better than writing for a newspaper, right?
204	P2	Even a university tries to make a distinction between scholarly activity and research.
205	P6	So let me try this … keeping the same example.
		[laughter]
206	P2	[overlap] So I wouldn't call it science, no.
207	P6	So the Rwandan genocide:
208	P2	[overlap] Yeah.
209	P6	[continued] Unique event; purely observational.
210	P2	[overlap] mm hmm
211	P6	[continued] As we try to explain what happened, we can test and falsify certain hypotheses.
212	P2	[overlap] mm hmm
213	P6	[continued] So we can sort of rule out some explanations as we go …
214	P1	[overlap] Based on other things.
215	P6	[continued] … because this exact same constellation of activities occurred and there was no genocide.
216	P?	[overlap] Right right right.
217	P6	[continued] So we can verify that this combination does not deterministically cause genocide.
218	P?	[overlap] mm hmm
219	P6	[continued] So we cannot explain the event, but we can rule out causal factors or causal combinations of factors …
220	P5	[overlap] In a {Popperian} sense, that's [can't hear].
221	P6	[continued] … in a hypothesis-testing framework. So does that then make it science?
222	P1	[overlap] I think it's science. I think it does. Social science, for sure.
223	P6	And if we can't do that, if there's absolutely nothing that we can do to start ruling out mechanisms, does it then fail the {facts}? And I ask this because I don't know where the boundary is myself.

(continued)

TABLE B.1

A Toolbox Dialogue Thread Extracted from DS1 That Addresses Issues Related to Scientific Methodology and Confirmation (Continued)

Speaking Turn No.	Participant	Speaking Turn
224	P2	[overlap] I think I wouldn't call it science. [can't hear]
		[mixed comments, talkover]
225	P6	[overlap] My sense is, [can't hear] … social science, you wouldn't?
226	P2	[overlap] Well I don't think … I mean, not all inquiry is science. Not all scholarly work is science. I mean, scholarly work around understanding love and romance and … in Shakespeare's time or even Shakespeare's view of the human condition, that's not science; it's {valuable} scholarly activity. [talkover] It could be research but it's not science.
227	P6	[overlap] Okay, so then if you want to draw a distinction between science and other scholarly pursuits, is scientific evidence privileged over other scholarly claims?
228	P2	It depends what you're applying it to. I couldn't apply … I don't think I could apply the rules of science to Shakespeare. They're not going to do—
229	P6	Yeah, but … so let's keep to the same example. So you've got … horrible things can happen to people …
230	P2	[overlap] Yeah. {They happen all the time, huh.}
231	P6	[continued] … So a volcano can explode; earthquakes can happen in central Africa—
232	P1	[overlap] Destroy the food supply.
233	P6	[continued] And so we know some things experimentally, and through credible simulation modeling, about how to prevent millions of people dying or hundreds of thousands of people dying {due} to some geophysical or {hydro} meteorological event.
234	P?	[overlap] mm hmm
235	P6	[continued] But there are also social phenomena that cause similar sorts of outcomes, which we can't study.
236	P?	[overlap] mm hmm
237	P6	[continued] So you want to tell policymakers where to spend money to prevent human catastrophe. So {does} the fact that we know something scientific about earthquakes and volcanic eruptions and floods mean we should spend our money preventing that, rather than on kind of softer … prevent ethnic tension?
238	P?	[overlap] Both.
239	P2	No, I don't think so. That's different, and that's not … that's a question that I don't think involves, necessarily, science. It really just involves human values and what it is we're after in the end.

TABLE B.1
A Toolbox Dialogue Thread Extracted from DS1 That Addresses Issues Related to Scientific Methodology and Confirmation (Continued)

Speaking Turn No.	Participant	Speaking Turn
240	P6	[overlap] Yeah, exactly.
241	P2	[continued] Are we after improving the human condition, or what? What is it we're after? And if the one component of worrying about the human condition and human safety had ... there is a scientific component to it, but there are also non-scientific components that have to do with human behavior as both individuals and as groups.
242	P6	My observation is that in policy circles, [talkover] the scientific evidence is privileged. And I say [this as] somebody benefits from that. Because you know, I deal in policy world and ...
243	P2	[overlap] Oh, it may well be. Whether I think it should be is another question.
244		[laughter]
245	P6	[continued] And sociologists and anthropologists, well sociologists hate it because we economists always win with the policymakers, because we {try} numbers. You know, and we can provide hard evidence
246	P?	[overlap] Yeah.
247	P6	[continued] that ... we often ...
248	P?	Statistical predictors {don't}—
249	P6	[continued] ... yeah, in the purely statistical sense,
250	P?	[overlap] yeah
251	P6	[continued] subject to all of our maintained hypotheses being true—
252	P1	[overlap] Highly uncertain.
253	P6	[continued] ... which is almost certainly not true, by the way ...
254	P?	[overlap] yeah yeah yeah
255	P6	[continued] and here's your evidence... We always win. [mixed comments, laughter, talkover]
256	P2	I get what you're saying, and I don't know whether I have a clear boundary in my head that would say "This is science and this isn't," or "This is worth pursuing and this isn't."
257	P5	[overlap] The really interesting thing—well, one interesting thing—is {when} you have a data set {of} about 50 species, milkweed species for example ...
258	P?	[overlap] That's a random—

(*continued*)

TABLE B.1

A Toolbox Dialogue Thread Extracted from DS1 That Addresses Issues Related to Scientific Methodology and Confirmation (Continued)

Speaking Turn No.	Participant	Speaking Turn
259	P2	[overlap] Why'd you pick those? Yeah … [laughter]
260	P5	In a way I imagine the … and this is sort of my own insecurity coming out, but I imagine that what I do with those data are very much like what a social scientist might do with data on the success and lack of success of different societies.
261	P2	Mm hmm.
262	P5	You're taking data on it, and running correlations, looking at the relationships. And so, certainly they … one interesting observation is, no one—I don't think—would say that what I do with those {phylogenetic} data are not science, maybe because the data set happens to be non-human related, I don't know why, even though the methodologies might be quite similar.
263	P1	So, I'm having trouble with [P2's] position here, because—
264	P2	[overlap] Well, explain my position. I'm not even sure what it is. [laughter]
265	P1	[continued] But I think it isn't … it is certainly not the physical or biological sciences that we were brought up in. And chemistry as well. Certainly not the way I was brought up. But I think it is definitely social science when you are doing this deductive reasoning. And, you know, I know you're not buying it, but—
266	P2	[overlap] Wait a minute, I'm not buying what?
267	P1	Well, you said you don't consider it science. You consider it scholarly work.
268	P2	Oh. I would say there are things that I would consider scholarly work and clearly not science. There is science, and there's a whole big grey area in between, where you'd have to think about, you know, I don't know, is {it} even important to know "Is it science or isn't it science?" If, as [name] says, things that have a label *science* on it tend to have more value or you can put some mathematical model to it or whatever, that you tend to get a better response, at least in the current society, for {plunking} or for risk factors or for whatever … [mixed comments, talkover]
269	P2	[continued] … or for the politics, then that question might matter. But I'm not sure, to me, just … I was going to say it and then take it back … I'm not sure it matters all that much to me. On the other hand, when I think about {our organization},

TABLE B.1

A Toolbox Dialogue Thread Extracted from DS1 That Addresses Issues Related to Scientific Methodology and Confirmation (Continued)

Speaking Turn No.	Participant	Speaking Turn
270	P?	[overlap] right
271	P2	[continued] I do think about … what I do think about is when we make investments, what's the impact it would likely have, and are we optimizing the investments we're making? That's a pretty open-ended question; I don't know. And I don't think all of it's science.

B.3 THREAD FROM DS2

The thread in Table B.2 is associated with Excerpt 10.1 from Chapter 10. The participants were members of an Integrative Graduate Education and Research Traineeship (IGERT) project created to work on biofuels, plant-made products, and environmental sustainability. The primary discipline for each participant (P) was as follows: P1 (engineering), P2 (engineering), P3 (chemical engineering), and P4 (microbiology). They discussed two prompts from the *Motivation* module:

MOTIVATION 3: My disciplinary research primarily addresses basic questions.
MOTIVATION 4: The importance of our project stems from its applied aspects.

TABLE B.2

A Toolbox Dialogue Thread Extracted from DS2 That Addresses Issues Related to One's Motivation to Practice Science

Speaking Turn No.	Participant	Speaking Turn
26	P3	How about number 3?
		[Pause]
27	P4	I feel like mine, my just general research in immunology is really looking at basic questions. And that's why I like the project I'm working on with you all because it is more applied, and I think that makes it kind of more interesting, but I don't know what you think of your own research.
28	P3	Yeah, I think it is applied but on the other hand, since my disciplinary research primarily addresses basic questions, so …. Umm, so that's where I was sort of neutral.
29	P4	[overlap] The primarily
30	P3	[continued] Yeah, the primarily … I mean our, my disciplinary research … I think it's motivated by the application, but I think it still addresses basic questions.

(continued)

TABLE B.2
A Toolbox Dialogue Thread Extracted from DS2 That Addresses Issues Related to One's Motivation to Practice Science (Continued)

Speaking Turn No.	Participant	Speaking Turn
31	P1	So is … this kind of research, this project? Or is it things outside of this one project?
32	P3	[overlap] I wasn't sure.
33	P1	[overlap] That is what I wasn't sure of either when I tried to answer that question.
34	P4	I interpreted it as being kind of, what you do without … you know, not in the group, not when I work with you, but what I do when.
35	P3	That's a good point. Yeah.
36	P1	Because the other thing is, I'm thinking about our project itself, that we're working on together and whether we would consider that to be applied or basic. Are we trying to answer an applied or basic questions or is it somewhere in the middle and that …
37	P4	Well I guess I think of it as being applied, but I guess in a way it's still basic. But compared to the other research I do, it's definitely a lot more applied.
38	P1	So it's not quite fundamental … because people have already studied some parts of the project we're working on. But it's umm … not quite basic. Or did I say that wrong? It's in between.
39	P3	Yeah. But … but see when I … when I look at it, it's motivated by an application. But there are a lot of fundamental questions.
40	P?	mm hmm
41	P3	[continued] How to design the genes, for example, that rely on, you know, studying the structure of best [can't hear] that's sort of … a basic question.
42	P4	[overlap] Uh-huh.
43	P3	[continued] Even though the application and the end goal, is more applied.
44	P1	Because in industry, when I was thinking about applied research, it was very soon. It was imminent that we were going to be doing this like next year or something. And umm, it's more on practically how do you go about doing it so that it works well, not can we do it, will it work sort of things.
45	P4	True. Kind of like the extreme of applied research is industry.

TABLE B.2

A Toolbox Dialogue Thread Extracted from DS2 That Addresses Issues Related to One's Motivation to Practice Science (Continued)

Speaking Turn No.	Participant	Speaking Turn
46	P3	[overlap] And it probably … and did it understand … did it address trying to get a fundamental understanding of what was going on? Or just get to the end goal.
47	P1	[overlap] Right there was already … right it was applied research and uhh … make a product.
48	P1	[continued] So the fundamental research had been done long ago, by someone else. And we didn't even get involved in that, we just got involved in the application of it. So that's why I wasn't quite sure if we're seeing this as something that would really be applied. You know, our project here or if it's something that could be applied as like a demonstration of it, but not necessarily this particular project being applied, actually produced this particular product that we're making. [unclear]
49	P4	[overlap] That's interesting.
50	P3	So you don't see it as being … as applied as …
51	P4	[overlap] yeah, it's completely relative to what we're used to doing. That is interesting … 'cause I'm used to just studying, you know, immune cells for the sake of knowing something about immune cells and then, you know, it's never necessarily, you know, you never even think of it as being able to go anywhere directly. So, this, to me, just seems more applied, but it's funny that … that's a good point.
52	P3	But you'd look at it as a … this research may end up in nothing.
53	P4	[overlap] It's basic.
54	P3	[continued] You know there maybe no … really useful outcome that will … will come from it. So that it's more … more basic.
55	P1	Cause that's a probability of success, you look at an industry. You know, what's the chance this is really going to work? And we haven't really talked about what are the chances that this would actually succeed and be able to go to that endpoint with this particular project.
56	P4	[overlap] Yeah.
57	P1	[continued] It is a different viewpoint from what I'm used to.

(continued)

TABLE B.2

A Toolbox Dialogue Thread Extracted from DS2 That Addresses Issues Related to One's Motivation to Practice Science (Continued)

Speaking Turn No.	Participant	Speaking Turn
58	P4	[overlap] Right. And I don't necessarily think that it will … will end up there. But to me it's like, oh well it has kind of potential to contribute to something that could …
59	P1	[overlap] Right.
60	P4	[continued] end up there that. That other parts of my research aren't at all … well I guess they all have some potential, but … there's more basic research than what I do, but … but it's certainly not as applied as this problem. That's interesting.
[Speaking turns 61–66 skipped—participants return to the thread in ST 67]		
67	P3	The importance of the project, did it stem from the applied aspects in your mind?
68	P4	Well, I either agree that there is importance to learning about the process, but ultimately, for the reason that you could apply it in other ways. Even just the knowledge I guess.
69	P3	[overlap] What did you say on that one?
70	P1	I was right in the middle on that one.
71	P2	I was neutral.
72	P3	Yeah, I was a little bit more on the agree side … because in some respects if you think this is something that will never actually … uhh … you know, end up being used, the outcomes of it then, the importance … uhh … I guess I should've said the importance should be from the basic standpoint right, from the fact that it will give us knowledge from the basic side of it.
73	P4	Huh. [Pause]
74	F	Do you have different ways of thinking about importance? Different ways of how you interpret that?
75	P4	That kind of comes back to is it important just to learn something or is it, does it need to be applied for it to be important. That sort of idea.
76	P2	Yes I think that the application is very important, so granted, our project is focused on actual application. But, you know, most research that breaks through usually comes from the [can't hear], so actually it's a basic, more fundamental thing, it's very important, but currently our project seems to be more focused on actual application. Yeah, so …

TABLE B.2

A Toolbox Dialogue Thread Extracted from DS2 That Addresses Issues Related to One's Motivation to Practice Science (Continued)

Speaking Turn No.	Participant	Speaking Turn
77	P3	I guess if you define, whether, you know, does importance mean the impact that the work is going to have? You know, does it impact you know other scientific researchers? Does it impact society? Does it impact um … yeah, so if you talk about impact, then I think it does stem from its applied aspects.
78	P1	I was thinking though that, the success or failure wouldn't be dependent on whether this could be really applied to something or not. So it's important even though it's not applied. It's in a different way though.
79	P4	Right, more to the scientific community than just advancing plant technology then.
		[Pause]
80	P1	Cause I think we learn from failures too. I mean, that's probably more in the basic type research than in applied research. But if things don't work, we still learn something and we might still learn something that is important from that.

B.4   THREAD FROM DS3

Table B.3 contains the thread associated with Excerpt 10.2 from Chapter 10. The participants were members of a university group interested in submitting a proposal to the National Science Foundation (NSF). The primary discipline for each participant (P) was as follows: P1 (natural resource economics), P2 (natural resources), P3 (natural resource economics), P4 (sociology), P5 (not listed), P6 (environmental engineering), P7 (engineering), P8 (ecology), P9 (engineering), P10 (fisheries biology), P11 (agriculture), and P12 (water engineering). They discussed a theme that did not correspond to a prompt.

TABLE B.3

A Toolbox Dialogue Thread Extracted from DS3 That Examines the Question, "Is Engineering a Science?"

Speaking Turn No.	Participant	Speaking Turn
402	P1	Can I ask a really silly question? Is engineering a science?
403	P4	Good one.
404	P?	[overlap] I think that …

(continued)

TABLE B.3
A Toolbox Dialogue Thread Extracted from DS3 That Examines the Question,
"Is Engineering a Science?" (Continued)

Speaking Turn No.	Participant	Speaking Turn
405	P7	[overlap] Well go to AGU [American Geophysical Union] and count the number of engineers there are there talking about science. You will be overwhelmed that so many engineers spend a lot of time talking about science.
406	P6	I think there's a lot of us particularly in academia and in environmental and in water resources where there's really, well I describe myself, I have the title of an engineer, but I'm no different than most watershed scientists. I mean, I do things a little bit different because of my background, but I overlap significantly. I could probably, I could be in watershed sciences. So really there's a subset of us that really are just scientists with a different title.
407	P1	But as people, but then when you think of how they're, they evolved, what their frames of mind are and how they're treated institutionally. Would you say that there's a difference?
408	P?	[overlap] Yes.
409	P?	[overlap] Yes.
410	P1	[continued] Because I think it's interesting that in this particular interdisciplinary group, there's not just the ecol…, the people who don't deal with people and people who do, there's people like engineers, and …
411	P7	[overlap] I would say in our department, in our academic reward system, our head and our dean, science is treated far differently than say, ok—so if I published a scientific paper versus an engineering paper—that's different. And I've experienced this with colleagues all over the country who are in engineering departments, but are teaching things like hydrology and other classes that have a lot to do with science, but their department heads want them to publish in {ASCE} [American Society of Civil Engineers]
412	P?	[overlap] ASCE
413	P7	[continued] journals, you know engineering specific journals when what they're really doing in their research is applied science, it's not engineering.
414	P1	But you made the distinction by saying, "do I publish here, or do I publish there?"
415	P7	[overlap] When I write papers I almost unconsciously write things like engineers and scientists do things like this. [talkover] So almost unconsciously I distinguish between the two, but I think it's such an artificial division. Like it's really hard to draw lines [cut off]

TABLE B.3

A Toolbox Dialogue Thread Extracted from DS3 That Examines the Question, "Is Engineering a Science?" (Continued)

Speaking Turn No.	Participant	Speaking Turn
416	P9	[overlap] The way I see division and I hadn't really thought about it until you asked the question [name], is that in the university engineering is a science, but once you leave the university, engineering is not science, it's a practitioner field.
417	P?	[overlap] Yeah
418	P9	[continued] And I really struggle with this with say upper division undergraduate students who are trying to figure out if they want to go to grad school or if they're trying to get out so they can get their engineering consulting job or state job or whatever. Because it's really different skill sets that they're, that, they're, that the students are interested in and that are good for them. That's how I see the difference.
419	P6	Yeah undergraduate preparation is not very scientific, but towards the end you become, there's more science toward the end in your particular discipline, I guess in some of the various disciplines within engineering. But once you get to graduate school I think it changes a lot from my perspective.
420	P9	I think that's true though of many, the way, the modern way we teach science is that way. People getting a bachelor's or even a master's in fisheries biology or in [name]'s department of ecology, they're not going to be scientists, they're going to go work for government agencies, or be consultants and collect data, assess, and monitor, or manage fisheries or manage water or build irrigation infrastructure and it's only—for the most part—the people who are going on with PhD programs, now they're doing science because they're doing what I described earlier I think as question-
421	P?	[overlap] asking questions
422	P9	[continued] driven, not hypothesis, but research, and I would completely agree with it. I consider all of you scientists, you know, but your graduates of your undergrad programs who are going to take the engineering exam and go work for DWR [Division of Wildlife Resources], well they're going to do engineering and that's different.
423	F	But can we go back, you, were you saying a few minutes ago that your performance criteria, there is a difference between how the administrators understand how you're characterizing what you do and how that's credited differently? Is that?

(continued)

TABLE B.3

A Toolbox Dialogue Thread Extracted from DS3 That Examines the Question, "Is Engineering a Science?" (Continued)

Speaking Turn No.	Participant	Speaking Turn
424	P7	Yeah definitely. Um, in engineering, in fact maybe I shouldn't be saying anything about my administration or I might get in trouble.
425	P?	P7 said …
426	P?	P7!
		[laughter]
427	P7	[continued] My impression of things that I've heard from faculty in my department is that my dean is very, he's mechanical engineering. He's not a say civil or environmental engineer who would deal with you know the physical sciences. He's very focused on patents and inventions, things that would apply to mechanical engineering but I'm never going to invent anything. [laughter] I might learn lots of really good and useful and interesting things to people, but I'm not going to apply for a patent. And so, the reward system being slanted towards that kind of a contribution is never going to work for me.
428	F	Um, so let me ask you two with a follow up. Does this affect or does this have any bearing on team science, with respect to what career stage you're at and how you make decisions?
429	P2	[overlap] It's almost like there's kind of a mixed signal that we get from our administrators. Like they say you need to be thinking about building collaborations and having an internationally recognized research program and working with other people. But at the same time we're not going to reward you for any of that. We're going to reward you based on these stupid stodgy rules that we have, but you have to be doing this other stuff too. So I mean, I'm approaching this from the perspective of a research faculty where, you know, I don't have to worry about teaching and the academic side nearly as much as other people do, so, and I meet once a year with my administrators. So I figure as long as I'm publishing and doing good work, I'm probably going to be good [talkover, laughter] because they're not paying my salary anyway. Because they're not paying my salary anyway, so they can tell me what to do but it doesn't really matter because they're not paying my salary. But somebody who is actually being paid by the department might tell you a different story.

TABLE B.3
A Toolbox Dialogue Thread Extracted from DS3 That Examines the Question,
"Is Engineering a Science?" (Continued)

Speaking Turn No.	Participant	Speaking Turn
430	P6	I actually asked [name], one of the other senior faculty involved in this project, about that exact question because I said, "listen, here's the expectations that we live by, here's where we get told in terms of being involved in these big interdisciplinary {NSF} [National Science Foundation] kind of projects like this one," and they totally go against each other. You know, how do you do that? And he started sending emails up the food chain and basically they're like, yep we'd like our young faculty involved in these interdisciplinary projects but essentially they can't rule, they can't be that big of a part of their, because they still have to have x-million publications and interdisciplinary projects take more time.
431	P?	[overlap]
432	P6	[continued] And publications are not as fast, you just can't ask a simple question, write a paper, and get it done. So you can't [unclear]
433	P7	[overlap] When I work on a paper with [name] too, it's probably not going to go to an engineering journal.
434	P6	But, I [unclear] engineering journal [unclear]
435	P7	Yeah.
		[talkover, laughter]
436	P9	I think for me I agree with you on that. What I took away from the um, was the warnings I've been given. And I think that's really just that, interdisciplinary work is great, you can do cool things, it takes lots of time, its high risk. And …
437	P4	[overlap] There might be ways to, as I've seen it done, I'd worry less about that because the projects typically do have pieces you can spin off along the way
438	P?	[overlap] mm hm
439	P4	[continued] so it isn't all or nothing we're going to only publish things as a co-activity, but um …
440	P6	That's fine
441	P4	[continued] It's worth keeping in mind as we move forward.
442	P6	[overlap] The question about independent pieces versus the whole?
443	P4	[overlap] Right
444	P6	[overlap] That's a necessary evil.

(continued)

TABLE B.3

A Toolbox Dialogue Thread Extracted from DS3 That Examines the Question, "Is Engineering a Science?" (Continued)

Speaking Turn No.	Participant	Speaking Turn
445	P4	[overlap, unclear] There might be some component pieces along the way, that you know you're doing your best work in the places you want to do it, [unclear],
446	P?	[overlap] Right
447	P6	Where but lots of things are being paid attention to.
448	P1	Well let me ask the question a different way. Um Elinor Ostrom makes some distinction between frameworks, theories, and models in terms of levels of specificity, and in a lot of the NSF stuff, science is perceived to be theory testing. So does engineering have a low recognized body of theory to which you are um encouraged to be on the cutting edge of? Are you doing theory testing? Because my impression of so much of engineering is modeling, at that level of modeling.
449	P7	I think environmental engineering, water resources engineering is very different than say civil, mechanical,
450	P1	[overlap] uh huh
451	P7	[continued] … those kinds of disciplines. I mean, engineering is big, and the environmental and water resources area is inherently riddled with science questions and so uh particularly environmental is an interesting thing to look at because there's, in environmental engineering there's solid hazards waste management, there's ground water remediation, there's natural systems engineering, there's drinking water, so much fits in here and they basically teach us, you have to be you know like a jack of all of these trades because you're an engineer and you should know how to do all of this stuff.
452	P9	Which is totally against the code, I mean engineering code. [laughter]
453	P9	[unclear]
454	P7	I think the answer is it totally depends on whether you end up in the practition, or whether you end up in science. Because if you're a practitioner, particularly if you're working for a consulting firm at the end of the day you're trying to make money. Which means you're not going to be testing a lot of theories … [laughter]
455	P?	[overlap] You're going to be cranking things …

TABLE B.3

A Toolbox Dialogue Thread Extracted from DS3 That Examines the Question, "Is Engineering a Science?" (Continued)

Speaking Turn No.	Participant	Speaking Turn
456	P7	[continued] Yeah, you're going to be cranking things you, you're pulling the code manuals down off the shelf and saying ok I have to do this, this, this, and this and I'm done and I think it's going to take me this much and here's my contingency fee and all that kind of stuff and at the end of day, you made some money. But in engineering there's also trying to advance the way that we're approaching problems and that's where I think the science and the theory testing comes in and it's, it's, I think it's just as scientific as any other discipline.
457	P9	Yeah, I think for me the modeling aspect is actually one of really the easy ways to test a theory, right? Because if we want to look at this factor and this factor and compare the effects, is this factor important, right, that's a question I can build a model very quickly to simulate those things and compare the results and see did it work out?
458	P6	Very quickly?
459	P9	Well, [unclear]
		[laughter, talkover]
460	P9	But regardless of how long the modeling process takes, it's the tool that actually answers the question to do the research. It's not necessarily part of, sometimes it can be just because I like constructing things and building fancy software interfaces for stuff, but there's also the question, answering the question, as a tool to answer those questions.
461	P5	I think the core of engineering is a lot of physical science, period. Fundamentally. You know?
462	P6	[overlap] Everything is built on that.
463	P?	[overlap] yeah
464	P5	You know, if you put that in a model and if you apply it you know. So the field does have, just in different way, I guess it's so much more, engineering is so much more applied. Um.
465	P?	[overlap] mm hm
466	P9	It got its basis I heard in France in the late 1700s when they building um canals and toll ways and roads and they had to figure out how to do that and then where to build the roads and which were going to be economically efficient ones and so you had to combine these three different disciplines and that's what engineering became.

(continued)

TABLE B.3

A Toolbox Dialogue Thread Extracted from DS3 That Examines the Question, "Is Engineering a Science?" (Continued)

Speaking Turn No.	Participant	Speaking Turn
467	P5	So it's interdisciplinary?
468	P9	Yeah it's been somewhat interdisciplinary by um by happenstance. Or at least maybe that's how it started.

B.5   THREAD FROM DS4

The thread in Table B.4 is associated with Excerpt 10.6 from Chapter 10. The participants were a team of three graduate students who participated in an IGERT project. The workshop was part of the IGERT launch meeting, which was the first meeting for this team. The primary discipline for each participant (P) was as follows: P1 (forest ecology), P2 (hydrology), and P3 (spatial ecology). They discussed one prompt from the *Motivation* module:

> MOTIVATION 1: The principal value of research stems from the potential application of the knowledge gained.

TABLE B.4

A Toolbox Dialogue Thread Extracted from DS4 That Addresses Issues Related to One's Motivation to Practice Science

Speaking Turn No.	Participant	Speaking Turn
0	F	Now we have an hour and 37 minutes so. Again the light facilitation is not to micro manage you all but you all can start wherever you would like if something struck you within the Toolbox that you would like to discuss. If you opt not to then I will start facilitating you. Do you guys just want to go down the list of our answers together and see how they compare and contrast? That can start us on the base on one to contrast and see where we are coming from.
1	P2	Okay, um, number 1, core motivation. Does the principal value of research stem from its applicability for solving problems? The first question is principal value of research stems from the potential application of the knowledge gained? And I had some questions about the value of research—that depends on who it is valuing for so think about it, to me is it individual researcher we are talking about because they could have a value of interest as well as a greater goal or is it a societal value or is it the value of their elder colleagues who are also super interested in this

TABLE B.4

A Toolbox Dialogue Thread Extracted from DS4 That Addresses Issues Related to One's Motivation to Practice Science (Continued)

Speaking Turn No.	Participant	Speaking Turn
		subject and then also it mentions the potential application so no matter what the research or how vague it could always have a potential application. People could argue in the future maybe it doesn't now but it could in the future potentially if you put it with these six other steps have an application so my answer I put, I feel like I agree with the statement. I think research should have a value potential application for the knowledge gained and that's a value for society is what I was thinking and we should do research that is valuable for society. That's mine. What do you guys think?
2	P3	I also agreed at the level 4 mostly I believe in applied research and science but I left it at a 4 because I think some research has value intrinsically in and about its self just for the benefit and knowledge and if it isn't directly applied and oh sorry.
3	P1	I was going to say I ended up valuing it at a three so kind of in the middle because I think research can have very applied or good application to society but I would agree with P3 that sometimes the intrinsic value of research is really important as well just doing research for the knowledge that you gain from it just to get a better understanding of biologically what system, what the constraints on a system are or how a system works and maybe down the road it might have some sort of applied practical knowledge when you dive into a certain subject more thoroughly but sometimes I just think good just to understand things just the way that they are.
4	P3	Intrinsic value of knowledge. Just answered to what's applied. If you are applying it to your own knowledge.
5	P2	Personally I guess in my view there are so many things you could look at and to every individual there are different interests obviously but I feel like the goals of what I want to accomplish are how can I help out the most people possible so I am going to shape my research goals in what has the largest societal significance and try to detract from my own interests. I guess I feel selfish in that way when I think about research well what's really interesting to me because to me I can find an interesting project and find something that's super applied to people, if I can get that overlap then everyone is happy kind of thing. I don't know I guess that is why I put it as a 5 because I feel like everything that I am working on I want it to be really applied to something that is a lot bigger than myself a few colleagues in my specialty fields.

(*continued*)

TABLE B.4

A Toolbox Dialogue Thread Extracted from DS4 That Addresses Issues Related to One's Motivation to Practice Science (Continued)

Speaking Turn No.	Participant	Speaking Turn
6	P1	I take the standpoint that I think environmental concerns are important as much so as societal concerns and that's probably coming from my background and standpoint in ecology is that you know humans a lot of times maybe negatively affect the ecosystems that we work in and so what I am interested in is and I guess some ecologists do this is separating society and nature to some degree but I think at any rate, now humans affect pretty much every ecosystems so we got to put them in the picture so it's good to apply that knowledge to society.
7	P3	What's our time frame?
		[talking about time frame]
8	P2	Something that I have picked up on already is definitely your background you're coming from the and I just realized it was different between us but there is a lot of value you see in what the greater happiness of the ecological system and the different components and I was looking at a human, is it nature for the value of nature or nature for the value of men? So that is what I was trying to figure out and then being a human well I guess am I trying to go for goals here but I guess overall I believe that through creating a sustainable ecosystem that meets the goal and is valuable for the sake of nature that's also going to meet our goals for future for sustainability for human population we can come into balance with that but I guess I am looking at the human issues as the main focus because I want to see, I want to help out as many people as possible. Yeah I guess that's putting them above that ecological system but definitely what you're saying that they are connected. I mean I guess we have core values on that which we are supposed to be talking about.
		[laughter]
9	P3	I kind of have gone through an evolution in that regarding throughout my career. I started very much discipline on the ecological side but along the lines that I work at you just can't, there is no way of getting away from that the involvement of the human population and in my opinion that is where if you are going to have long-term effective change that is where the change really has to be based because of our influence.

B.6   THREAD FROM DS5

The thread in Table B.5 is associated with Excerpt 10.3 from Chapter 10. The participants were a team of four graduate students who were participating in an IGERT project. The workshop was part of the first meeting for this team. The primary discipline for each participant (P) was as follows: P1 (economics), P2 (ecology), P3 (entomology), and P4 (natural resources). They discussed three prompts from the *Values* module:

VALUES 3: Value-neutral scientific research is possible.
VALUES 4: Determining what constitutes acceptable validation of research data is a value issue.
VALUES 5: Allowing values to influence scientific research is advocacy.

TABLE B.5
A Toolbox Dialogue Thread Extracted from DS5 That Addresses Issues Related to Values in Science

Speaking Turn No.	Participant	Speaking Turn
337	P1	Value-neutral scientific research is possible. I put that I … a 4. I agree with that. I think that you can take yourself out of the equation and do research that is antithetical to your perspectives or something like that, but I would think that most people choose not to because it just doesn't make sense in my mind, but that it is possible to do it. I put that I did agree with it.
338	P3	I kind of had the complete opposite view. I think that it's not really possible. People strive to do that, but there's always going to be those values in your mind that frame everything you think about.
339	P1	Is there any research question, I guess, that doesn't necessarily impart a value? Does orange juice help prevent rust? There's not necessarily a value system that is behind that research question.
340	P3	What draws people into their specific field is personal. That's why when you do research, unless you're Mr. Wizard, those values will come into play.
341	P1	That's why I gave it a 4. I agree with you. I think that there's … I don't necessarily see that many people would choose to do research that's outside of their value system because they went the direction that they went for a reason, but that it is possible to do research that is outside of your value system.
342	P4	What did you put, P2?

(*continued*)

TABLE B.5
A Toolbox Dialogue Thread Extracted from DS5 That Addresses Issues Related to Values in Science (Continued)

Speaking Turn No.	Participant	Speaking Turn
343	P2	I had a hard time. I didn't remember … I put a 5. I agree. I'm having a hard time agreeing with myself.
344	P4	Maybe you mismarked it.
345	P2	I don't remember this section being difficult. Each one of these I'm having a hard time with now. Initially, I thought it was like I put agree, strongly agree. You can do value-neutral scientific research. It's not good research. I'm changing the question, really. It is possible. Maybe I should put like a 4 or something like that. I think people do it. I think it's rare. I don't think it's very useful.
346	P4	I put a 1. I don't think you could ever do any scientific research without having values play a role in some fashion. Maybe your value is to do really good science. Maybe that's your value, but that's still a value. Maybe your value is to not do good science. There's always values.
347	P1	The next question: determining what constitutes acceptable validation of research is a value issue. I don't remember what my thought process was on this, but I put it as a fairly neutral. I gave it a 3 because I think that you are going to place values. They are going to influence the research data and the validity that you think that it holds. I do want to give some value to that. I think that you can give validation to research that maybe is contrary to your values.
348	P3	I think I looked at that in a completely different way. The … whether the research is valid is a judgment call that you make or a value that you have. That was a completely different definition of the word "value." I think that's true. I think that at some point you have to decide whether or not what you're doing is valid, hopefully, before you start.
349	P4	I gave it a 5, as well. Someone's got to make a call to say is that acceptable. You got make a … that's got to be based upon a value.
350	P2	I gave it a 5, as well. I think similar reasons. I hear this language. I know it's not researchers trying to convince themselves. I've done my data collection. I've analyzed it. Now, I need to convince myself that is correct. That sounds like a validation method. I believe there is a place for consensus. You can say, "Hey, there's my results." Someone will come over and say, "No, it's wrong because of this." You get that dialogue. I think it is initially a value issue. You have to … if the numbers are more important than qualitative, those are value issues.

TABLE B.5

A Toolbox Dialogue Thread Extracted from DS5 That Addresses Issues Related to Values in Science (Continued)

Speaking Turn No.	Participant	Speaking Turn
351	P3	I think that allowing your values to affect your research is advocacy. I think this relates a lot to question 25, whether value-neutral research is possible. Y'all said it was. Ideally, you don't want to be an advocate. We have people on the side who are advocates. That's not why we're here. Although, you have your personal perspective, which is going to frame your research, which is fine, it's best if you can remove those opinions about what should be if you can.
352	P2	I was on the middle on that one. I don't think you should start your scientific research from the position of advocating. I think that's a bias right there. I think that's to be avoided. I also believe scientific sector has responsibility to say, "Hey, this is what we know. If you do this, here's the consequences. If you do that, here's the consequence." Some of those are so obvious that they equate to advocacy. Stop doing this because it's just bad. I'm kind of neutral on that.
353	P4	What if your value is to be very objective in your scientific research?
354	P2	Then I think you're an engineer. You're a professional. I really believe there's a reason for doing research, and it's not just to be objective.
355	P4	If your value is to do that and your research comes out and it says this, is it advocacy?
356	P2	That's where I'm neutral because I think you still have to have an end result and say, "This is what we found." People are always going to ask you, "What should we do about this problem?" You can't just back up and say, "Well, my results say dot, dot, dot," and make it real dry and objective.
357	P3	Right, but once you're at that point, the research is done. It's like then it's different.
358	P4	Here's something for you guys to think about on that. That may be what everybody says and that's what they want, you don't ever let your values lead to advocacy, but 99% of the time it does. Almost everyone who's a wildlife biologist or soil scientist advocates for their field of study through some method. They start studying wolves. Wolves are the best thing ever. Don't shoot them. Don't mess with them. Plants, don't be grazing them. Their scientific research already is being influenced by that. It's the frameworks in their head. Now, if you had a guy on the other end of the spectrum,

(continued)

TABLE B.5

A Toolbox Dialogue Thread Extracted from DS5 That Addresses Issues Related to Values in Science (Continued)

Speaking Turn No.	Participant	Speaking Turn
		I hate wolves. I'm going to go study wolves to prove how they're not good. The framework's already in their head so … I put I disagree because I think all scientific research is influenced by values. That almost always leads to advocacy or I guess always does lead to advocacy in some fashion. If it's not you doing the advocacy, it's someone else doing the advocacy using your data, saying, "Well, look at this research."
359	P2	I've got a question for everybody. Do you think advocacy is a negative term?
360	P4	It's used as such, yeah. It is. I think most people consider it as bad.
361	P3	I think that if you're really a vocal advocate and you're researching a controversial topic, then your research is going to seem less valid. I think it's easy to question the objectivity regardless of how careful you really are when you're doing your research.
362	P2	So it's your degree of advocacy? When do you leave enthusiasm and arrive at advocacy?
363	P3	That's a good question. I think that depends on the exposure you have to … what exposure other people have to you as an advocate. That's not a question I can answer.
364	P1	I put pretty neutral on it. I guess as kind of a counter argument to P4, an example I have is I'm personal friends with a wildlife biologist who works for {state} Fish and Wildlife. He got into it because he loves animals. He loves the research associated with it. He does get to work on projects that he is passionate about. That's what got him into that. However, there are other projects that he's mandated to do by his supervisor upon which he disagrees with him. He still has to take an objective, scientific approach to doing that. He's not an advocate for that particular research or what the outcome of that research is going to be. He may even disagree with it, but he's still doing that research, trying to approach it from an objective way. In addition to that, working for a government entity, he's required to take a neutral stance politically. He cannot advocate. Even though some of the research you do may support or not support some of his personal values, he's not personally allowed to advocate for those things.
365	P3	It depends on how you define "advocate."

TABLE B.5

A Toolbox Dialogue Thread Extracted from DS5 That Addresses Issues Related to Values in Science (Continued)

Speaking Turn No.	Participant	Speaking Turn
366	P2	Do you think his private advocacy influences scientific research? It sounded like it you were saying he wasn't because he does these different tasks that he doesn't … he's not a fan of.
367	P1	Right, and I mean he works on a team that I think helps to ensure his objectivity. He's not the only biologist that is like "I don't like what this data point is telling me. I'm just going to erase this data point right here." He's on a team of people that helps ensure the objectivity happens. He still has to do research that he ultimately knows is probably going to lead to policy decisions.
368	P2	Advocacy is not falsifying data.
369	P1	He's doing research in an objective fashion that's against his value system, though. It was just kind of a counter argument to P4 that he's saying that you cannot separate values and advocacy in research. I'm saying, "Well, I disagree with that." I think that you can. I have seen it in the past where it's been not optional for people because of their positions. I mean, he's OK with it because he still gets to do some of the research that he's passionate about. He also … he has that other side of that coin that he's responsible for. He works equally hard on both because of the pride that he has in his job and wants to keep his job and all that stuff.

B.7 THREAD FROM DS6

Table B.6 contains the thread associated with Excerpt 10.5 from Chapter 10. The participants were attendees at a workshop aimed at developing and sustaining interdisciplinary graduate programs. The primary discipline for each participant (P) was as follows: P1 (engineering), P2 (qualitative research), P3 (qualitative research), P4 (genetics), P5 (communication), P6 (family therapy), P7 (family science), P8 (anthropology), P9 (health sciences), and P10 (psychology). They discussed four prompts from the *Reality* module:

REALITY 1: Scientific research aims to identify facts about a world independent of the investigators.
REALITY 2: Scientific claims need not represent objective reality to be useful.
REALITY 3: Models invariably produce a distorted view of objective reality.
REALITY 4: The subject of my research is a human construction.

TABLE B.6
A Toolbox Dialogue Thread Extracted from DS6 That Addresses Issues Related to Science and the Nature of Reality

Speaking Turn No.	Participant	Speaking Turn
197	F	So what about, uh, 2 on that sheet, on that page in the module, reality module? Um, well, actually 1 and 2. Both seem to be related to … to maybe to the discussion we just had about models and … but one earlier to the point about … to the conversation about values. Right? Like you said, objective reality was a part of the focus in your discussion about the value section and that's related … it's to the first of the prompts. I'm curious if there are other thoughts about that, those—those two prompts.
198	P10	I'm sure like you know scientific claims need not represent objective reality to be useful and so are useful to who? [laughter]
199	P5	I had the same issue.
200	P10	[crosstalk] It can be useful whether it represents objective reality or not as long as it advances your agenda. [laughter] But if you're … or if you're coming from a scientific background you might take a different viewpoint on that. So, uh, yeah, I think a lot of people have made a lot of new pseudo-scientific claims that didn't turn out to be represented in reality very much.
201	P4	Such as the presidential debate. [laughter, crosstalk]
202	P9	Or scientists might use that to advance their own funding agenda, right?
203	P5	[crosstalk] I circled that one, um, because I know, uh, some disciplines like to, uh, debate non-realities for the sake of higher understanding, but I did not circle my response, uh, with respect to that, but I answered it from a human behavior perspective, so we're going to be, be able to understand one person, there seems to be some sort of value.
204	F	So you put a 1?
205	P5	I did.
206	F	From your perspective. So could you articulate that?
207	P5	Apparently I didn't do that very well at all. [laughter] Give me a minute. Somebody else talk and I'll be back.
208	P4	So I gave it a 2 because it's fairly common in science to make approximations. To simplify the problem you hope those approximations don't matter. They were sort of ancillary, there might be sort of boundary conditions for the problem and they won't hopefully have too much of an effect.

TABLE B.6

A Toolbox Dialogue Thread Extracted from DS6 That Addresses Issues Related to Science and the Nature of Reality (Continued)

Speaking Turn No.	Participant	Speaking Turn
		And then you get a robust result. So there could be a practicality keeping it simple so that the bridge engineer gets it right most of the time.
209	P1	And you don't need every detail to build a bridge. You can ignore that animistic nature of, of concrete.
210	P4	Right.
211	P3	The thing is if you're thinking about bridges then I might have answered all of these differently.
212	P1	That comes down to a lot of its perspective [crosstalk] …
213	P1	… in interdisciplinary research.
214	P5	Yeah.
215	P1	I always think of that [unclear] [laughter].
216	P2	Say, number 1, um, number 1 says scientific research aims to identify fact. Well, there is … I would claim … I would argue that there is scientific research that does things, but it's not identifying facts. So there's … it's not driven by a search about facts, it's driven by a search for understanding human meaning or human experience, which is a different sort of research.
217	P5	Or process.
218	P2	Yeah, social processes.
219	P1	So would you take the word "fact" and replace it with "knowledge?"
220	P2	Well, I said agree, 5, and said some researchers argue this but I'm not one of them. So I don't agree … I don't do that kind of research.
221	P3	I circled 1. [laughter] I mean, why would I argue?
222	P2	Um, but I'm okay with people searching for facts because I want to drive over a bridge that's going to stay in place [laughter] but, right, if you're doing other kinds of research maybe that's not your work and I think the two kinds of work … different kinds of work from this spectrum can co-exist, hopefully peacefully.
223	P1	This is like part of the scientific analysis has some predictive power. Like certainly in physics and biology like once you learn something about something you should be able to make predictions about what's going to happen again. Isn't that true in the social sciences as well?

(continued)

TABLE B.6
A Toolbox Dialogue Thread Extracted from DS6 That Addresses Issues Related to Science and the Nature of Reality (Continued)

Speaking Turn No.	Participant	Speaking Turn
224	P7	But process—but process research can still be predictive. You can understand what's—where something is going because of predictive value but your understanding the mechanism through which it's getting there which isn't necessarily a fact.
225	P1?	Well that's a fact. Mechanisms are …
226	P7?	That's not a fact. [laughter]
227	P5	Key different …
228	P7	It's not a fact.
229	P7	I used the wrong word talking to you. Used "mechanism," but seriously, I mean, different language.
230	P6	But as a researcher in family therapy, at times research is done as a …
231	P7	Right.
232	P6	It's to emancipate, transform. Uh, so there may … I mean, that could be one is to predict, so that does exist, but there are other alternatives as well what one might be doing it for.
233	P1	What do you mean by that? Emancipate? Like a particular person?
234	P6	No, active in the community.
235	P2	Well, school research is a really good example of this, because these systems where you have many … so many things that are impossible to measure and very complex systems that it's … like I don't know how you do predictive, but obviously there's a push at federally to do randomized clinical trials in schools but it's so impossible to do that … some of that work is not very meaningful. So maybe we can like trace social change or we can look at things that are happening in some kind of context and describe it and understand it better. But whether we can ever predict, uh, like where … [P3] and I are working on a project right now which I think the contractors are trying to predict, to predict what qualities might affect their school leaders. I am constantly frustrated with that.
236	P6	Going back here. So let's say there's a study on racism in a school district, its—and its participatory actions, and you're involving participants in the study itself, which is probably a very different practice—methodological practice. But it's to make a change in the system as well as to learn about the system.
237	P1	Right, but so then that's two things, you're learning about the system, that's the science part.

TABLE B.6
A Toolbox Dialogue Thread Extracted from DS6 That Addresses Issues Related to Science and the Nature of Reality (Continued)

Speaking Turn No.	Participant	Speaking Turn
238	P6	And but it's not nec … moving just away from predicting, so it's, there's, and … it's not necessarily trying to predict, it's to understand and make a change. So that's just a different kind of prediction.
239	P9	Could I throw something in because I kind of sit in the middle of the methodology here. Is it maybe what you're trying to say in the hard sciences, scientists like the biological sciences for scientists, for example, is that you get the knowledge and that starts building, right? So there's like … over time, it keeps building. Even though they're not necessarily trying to predict with their work, your work still does contribute to the body of knowledge which then people continue to keep building. It's the same way. It's just not quite as … not like bridges …
240	P5	So …
241	P9	… linear maybe.
242	P5	Yeah, it could [crosstalk].
243	P1	[crosstalk] big bridges, maybe.
244	P9	Exactly.
245	P1	Yeah I mean I think … you know, if you're doing a specific project where you're trying to not just learn about racism in effect, but also make a change, I mean the part where you're … the aspect of where you're trying to make the change I mean that's … that's almost more of an engineering type problem like where you're actually trying to do something.
246	P9	Applying the knowledge.
247	P1	Apply the knowledge, right.
248	P5	Can I ask you a question about objective reality though? [laughter]
249	P9	Sure.
250	P5	That's my poorly stated question. No, but, uh … because I think maybe when I read "objective reality," I'm reading it differently than …
251	P3	We all are to some …
252	P5	I think we all are … that's my … yes, that's also … let that point be known that … so if I'm studying racism in schools, what's the objective reality? Is the objective reality the, uh, general consensus? Is it one particular person going to that school? I think that's why I was also struggling with that

(continued)

TABLE B.6
A Toolbox Dialogue Thread Extracted from DS6 That Addresses Issues Related to Science and the Nature of Reality (Continued)

Speaking Turn No.	Participant	Speaking Turn
		question. Because I spend a lot more time just looking at individuals sometimes from the objective reality ... reality standpoint versus the objective reality of the world which with humans that's difficult versus studying, I don't know ... does an atom have an objective reality? So I think again there's multiple levels of this conversation.
253	P7	Please start talking unit of analysis.
254	P5	Basically.
255	P3	Yeah.
256	P7	So it's making the conversation at least for me complex because I don't really know how to answer where the objective reality is without thinking about making a scientific claim.
257	F	How, how .... Would you be willing to try to maybe define ... give it a shot, with like ... articulate the reason? One of these nuggets that you've I mean you've given examples and that's great. I'm curious if, if ... it's a great question that you've asked and I'm curious if you might be willing to take a shot at ...
258	P5	Could I ask my colleagues to help me out here? [crosstalk, laughter]
259	P7	Lifeline?
260	P4	A lifeline. [laughter]
261	P5	Somebody ... somebody who [crosstalk] crazy lady or actually getting, you know ... [crosstalk]
262	P10	I would add to that because when I did my dissertation I studied existential phenomenology which is the only philosophy I've been exposed to a bit. [laughter] But in that ... but in that philosophy they distinguish between three different types ... three different worlds, basically what they call the umwelt, which was the physical world, the midwelt, which was the interpersonal world and then the eigenwelt, which was your internal world. And so since then I've sort of thought about it as three separate realities, you know, that depends on which space you're operating in as to what methods and perspective you take.
263	P3	What about the lebenswelt?
264	P10	Which one?
265	P3	The lebenswelt?

TABLE B.6
A Toolbox Dialogue Thread Extracted from DS6 That Addresses Issues Related to Science and the Nature of Reality (Continued)

Speaking Turn No.	Participant	Speaking Turn
266	P10	They didn't teach me about that one. [laughter] I'm missing the fourth level.
267	P3	Well, the lebenswelt is actually the interaction between the ...
268	P10	The three?
269	P3	Yeah.
270	P10	Okay. Lebenswelt. Thank you. [unclear] Cool, but it helps me to think about shifting perspectives like that.
271	P6	Well, Jung does that too, right?
272	P6	and [unclear].
273	P10	Yeah. So he ... very influenced by [unclear], so.
273a	P6	It's kind of a post-modern viewpoint, but there's no one reality.
274	P10	But it's ... it comes more down to [unclear] kind of whatever units of space you're in.
275	P5	You're not happy with that though, right? [laughter]
276	F	I'm happy as a clam.
277	P5	I know you're just solitaire over there.
278	F	I'm just curious. [crosstalk] Does 4 on that page help at all to kind of ... it ... maybe?
279	P4	Which page are we on?
280	Several	To say that everybody has their own objective reality.
281		[crosstalk]
282	P5	Well, we could say that and I actually ... yeah, you could say that and I actually don't ... that's not my personal ... the way I run my research world, but yes I will acknowledge that, absolutely.
283	P1	[unclear] tying them together.
284	P5	Exactly, and so I say impersonal, you know. [crosstalk]
285	P3	Okay.
286	P5	And I study ... you know, I come from a much more, uh, indivi—individualistic perspective and I argue that we're all strategic, we—that's all controversial, right? But the idea of an objective reality, uh, with regard to human behavior ... and I think that's why I screwed up that question.
287		[crosstalk].
288	P7	No, no I'm sorry, there's no right or wrong answer.
289	P5	Correct, but when I was ...

(continued)

TABLE B.6

A Toolbox Dialogue Thread Extracted from DS6 That Addresses Issues Related to Science and the Nature of Reality (Continued)

Speaking Turn No.	Participant	Speaking Turn
290	P7	[crosstalk] I was just going to say I don't think you screwed it up in the sense that it …
291	P5	I didn't screw it up, but … [laughter]
292	P7	In the sense that …
293	P6	It's her objective reality, don't mess with it.
294	P7	I'm sorry, yes, I'm wrong. I revise.
295	P9	But it starts with a conversation about what the heck is it that we're talking about. I mean what is the definition of "objective reality?" If you're going to work on a team with people who don't share your opinion, you at least have to have the conversation about how everybody is constructing that. So you didn't screw it up because we had a conversation.
296	P1	[crosstalk] jargon and interpretation of the key terms, right?
297	P5	[crosstalk] But I think some of the issue is though is the key terms. There are some terms that are useful to have as starting points then to move away from.
298	P6	Well, this takes us back to philosophy of language, which you might be interested in. Uh, I mean, hermeneutics is the … the terms we're using because we've situated disciplinary, politically our own cultural background, all of that. So this is where I think we can get some of the same answers even though we're talking about it very differently. It's amazing we communicate as well as we do, actually.
299	P3	That's because … that's hermeneutics helping us.
300	P5	It's the miscommunication that brings us together. [laughter]
301	P3	So the thing that you were struggling was you were trying to match objective reality with the questions you were asking.
302	P5	Oh, sure. Absolutely.
303	P3	Which are basically from a—from a, um, a positivist perspective, it's subjective reality that you're talking about and so you couldn't answer the question because you're suddenly faced with that dilemma, which is to talk about cultural construction of racism or the systemic—systemic construction of racism from an objective reality perspective. Um, now people do it, I mean, the critical … a lot of critical social scientists are, um, critical realists and they accept racism as objective reality … that's the objective reality that exists and everything works from there. So again, it's … being that it's a question of ontology and it's …

TABLE B.6

A Toolbox Dialogue Thread Extracted from DS6 That Addresses Issues Related to Science and the Nature of Reality (Continued)

Speaking Turn No.	Participant	Speaking Turn
304	P9	And objective reality is the exact tension between historically the hard sciences and the soft sciences. The hard scientists say, "I measured it, here, I did the measurement, it's here." And the soft scientists, when they measure something, "How do you measure construct you can't see or feel or ...." Right? So there is the tension.
305	P3	And that's why I thought the model question was really the best question.
306	F	The model question?
307	P3	The [crosstalk] model question you know because [crosstalk].
308	P5	Three.
309	F	So that, number ... models invariably produce ...?
310	P3	Yeah, because it really forces us to look at the measurement and look at what we're talking about [crosstalk] so the relationship of research questions to reality. So, good question.

Index

Page numbers in *italic* denote figures and in **bold** indicate tables.

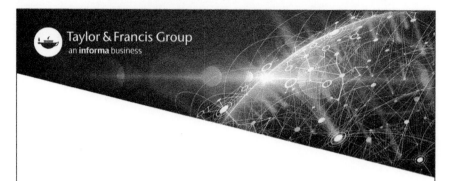

Printed and bound by CPI Group (UK) Ltd, Croydon, CR0 4YY

21/10/2024

01777058-0006